NOTHING
TOO GOOD FOR
A COWBOY

ALSO BY RICHMOND P. HOBSON, JR.

Grass Beyond the Mountains
The Rancher Takes a Wife

RICHMOND P. HOBSON, JR.

NOTHING TOO GOOD FOR A COWBOY

A TRUE ACCOUNT OF LIFE ON THE LAST GREAT CATTLE FRONTIER

SEAL BOOKS

Seal Books and colophon are registered trademarks of
Random House of Canada Limited

NOTHING TOO GOOD FOR A COWBOY
Seal Books/published by arrangement with
McClelland & Stewart
McClelland & Stewart edition published 1978
Seal Books edition published 2015

Library and Archives of Canada Cataloguing in Publication
data is available upon request.

ISBN: 978-1-4000-2663-0

Cover design by Five Seventeen
Cover photo: © Tetra Images / Getty Images

Printed and bound in the USA

Published in Canada by Seal Books,
a division of Random House of Canada Limited

www.penguinrandomhouse.ca

10 9 8 7 6 5 4 3 2 1

Penguin
Random
House

This book is dedicated to GLORIA

CONTENTS

CHAPTER I

The Decision

It was October, 1939. George Pennoyer, iron-jawed boss and general manager of the Frontier Cattle Company, Limited, whose four million acres of controlled range land sprawled across the wild, little-known interior of British Columbia, faced Panhandle Phillips and me in our ground-floor room in the historic Quesnel Hotel, in the Cariboo district of British Columbia. Our boss's face was grave and hard, expressionless to an outsider, but to me, who had known and worked for him, and admired him for so many years, the telltale crinkles spreading like little fans from the corners of his icy blue eyes revealed an unaccustomed sadness.

I shall never forget our boss's words; his sum-up of Pan's and my five years with the cattle company. I realized that a major crisis in our lives was at hand.

"Sit down, boys," George said. "You might as well make yourselves comfortable while I round this situation up for you."

Pan and I shuffled about, neither of us finding a suitable place to sit. Pennoyer came straight to the point.

"Boys," he said, "the cattle company has blown its cap—the layout you buccaroos have thrown yourselves

into has exploded like a car of stumping powder—"
Pennoyer paused a minute and pointed to the old-
fashioned spring bed in the middle of the room. "Grab
a seat on that bed. When I get through with you, you'll
need a good comfortable place where you can lie down
and cry."

I moved hurriedly to the bed, awkwardly stepping
on the protruding paw of my trail partner, the Bear, a
mongrel dog who had thrown in with me several years
before.

Pan straddled the brass foot rail, his deeply tanned,
long-jawed, Roman-nosed face serious.

"Shoot, George," Pan said. "I don't think you're
going to tell us anything we don't already know, but we
want to hear how you're going to cut out the culls."

George lit his old, life-worn pipe. Instinctively he
shielded the bowl for a moment with his rope-calloused
hand. I noticed his steely blue eyes narrowing to slits. He
stared at the toe of his riding boot.

Pennoyer said, "The financial backers of the company
were making sense on the telephone this morning. They
have the right idea when they say that from now on and
to the end of the war, it will be impossible to line up
enough top hands to carry on with the Frontier layout.
It's October now. Our hay crews have left us right in
the middle of haying to join up. The war has upset the
directors' plans and they can't guarantee us any further
financial support.

"We've got a smart bunch of purebred cows back
there on the range, worth a lot of money to us as breeding
stock, but very little as beef. We're going to have to sell a

big bunch of those critters to pay off the company debts."

"Whoa there a minute," exclaimed Pan. "You're going a little too fast for me, George." Pan snapped to his feet and then ducked under the bed for the case of beer he had hidden there the night before. He slid the beer case out and, snapping off the tops, passed around bottles.

"You'll need it," Pennoyer drawled, "to fog up your minds."

I felt the tension in the room recede and leaned back on my elbow, took a long gulp of beer, and looked at Pennoyer. Pan restraddled the foot rail of the bed.

George said, "Both of you boys are foremen of a layout with great possibilities, the granddaddy of controlled land areas, and I know what you've put into its foundation during the last five years, and what its success means to you both. But the cards just aren't stacked in your favor.

"Looking at this setup from a purely practical point of view, you boys, with what little help you've got—a few Indians, a cowhand or two, and a couple of greenhorns from the city—will be marooned back there on the most isolated, the most remote cattle ranch in North America.

"You're back there two hundred miles by pack horse and wagon trail from Quesnel here, the nearest town. You'll be cut off from any contact with the outside world from freeze-up in November to break-up in May—six months of the year. You haven't got half enough hay to winter the cattle that we trailed in there. You haven't got the men to plant at the hay units to feed the cattle this coming winter even if we did have the feed."

George paused for a moment, taking a gulp from his bottle of beer. Pan and I remained silent.

Pennoyer continued: "There're the cattle drives, moving herds from unit to unit through dense jackpine jungles where, as yet, few trails are cut. You've got to have top men—good riders, men who can handle cattle in rough, dangerous country. As you fellows have already found out, it's a rough deal when a herd breaks or stampedes on you in these bushes, not like the open country in the States.

"When the company was formed and we had money and men to open up this new country and plant cows on the range, there was nothing to it. But, now that the war has hit, the money and the men have flown away."

George cleared his throat.

"To sum up," he said, "we haven't got the men, the money for this size operation, or the backing, and if you boys tried to carry on you'd be risking not only what capital the financiers can still get out of the sale of assets— but your own lives as well."

George relit his pipe. Again he looked grimly at the toe of his boot.

"Then you figure the only sensible, businesslike thing to do is close down now?" I asked him.

George nodded his head.

Pan was staring across the room at Pennoyer. He snorted loudly, took a careful drag at his tailormade, and blew smoke through his long nose. He spoke in a twangy, nasal voice.

"George," he said, "us fellows know you've got to go back to Wyoming to your family, your own ranch and your businesses in Lander. But suppose you was in our place right now—thirty years old. There is gonna be a shortage of beef and the Frontier is waitin' back there with cows already belly-deep in grass. Suppose you figured there was a fightin' chance. What would you do yourself?"

George looked sharply at Pan. His eyes glittered like newly polished steel.

"What would I do?" he asked. "Why, hell I'd take the big gamble—I'd stick with the layout. I'd tie hard and fast and throw my rope in there just as far as I could snap it."

Pan grinned widely. "And that's just exactly what Rich here and me is gonna do."

The Bear's tail thumped the floor.

So Pan and I threw our hats in the ring. But we were faced with certain important problems that had to be solved before we saddled up and rode out of Quesnel.

First, there was that universal stumbling block—money. Pan and I went through our saddle pockets and pack boxes collecting our books. These quite-unorthodox records consisted of notations on brown wrapping paper, figures written on the back of cigarette-paper books, and small pocket pads, blotched and warped by rain, sweat and river water.

George shuddered, looking at the piled-up mess on the bed.

"If the tax inspectors and government accountants see that," he said, "both you boys will be hung. With a little training that dog under there could do a better job of bookkeeping."

He stepped to a desk in the corner of the room and returned with a new, schoolboy-composition book. Sitting on the edge of the bed, he began copying down our figures under Assets, Liabilities, Accounts Receivable, Bills Payable, and so forth. On the accounts-receivable end there were few notations, and our bank balance was alarmingly small.

On the payables end there were many figures to be jotted down.

Some of our notations were hard to understand.

"I can't make heads or tails out of a lot of this stuff," said George. "Here's Rich's writing on the back of this wore-out old bank check. I can't read it."

He squinted closely at the memorandum, cleared his throat and read in an incredulous voice: " 'One gallon 30 O.P. rum, ten-year-old gray mare with slight string halt named Peony, and fifty dollars for McNutt Ranch. Payable to old McNutt, October first.' "

"That's for the McNutt place," I said happily. "Three hundred and twenty acres—on the trail to Quesnel. Cabins, barns, good holding ground and fall rustling for the Batnuni bunch."

Pan snorted. "I got loused up in that cabin once." He spoke in an irritated voice. "Had to burn up a damn good suit of Red Label underwear, and old Willie-Be-Good snagged hisself in the brisket on one of those crazy manger poles in that barn."

George shook his head.

"Did you get the deed to the property?" he asked me.

"Sure—it's in the bank here in town. I pick it up when I deliver the money and Peony. I gave old McNutt the rum as a down payment."

"That's a lot of money to pay for a place like that," said Pan. "I broke through that bridge at McNutt Crik. Got the outfit into the mud to their backs and damn near drowned myself and old Baldy."

George started to laugh. Pan grinned.

"O.K.," said George. "McNutt place goes down, as three-hundred-and-twenty-acre ranch with modern buildings and bridges, on the asset side."

He wrote it down in the ledger. He bit the end of his pencil.

"Now, we mark down on the accounts payable: Willie-Be-Good and fifty dollars."

Pan interrupted: "Not Willie-Be-Good. He's stove up back at the Home Ranch. It's Peony that Rich traded."

"Here's some of Pan's cancelled checks on the Canadian Bank of Commerce," said George, taking up a moldy packet. "We better note them down on the books now."

He cut the moosehide wrapping with his stock knife and examined the notations scrawled on the lower left-hand corners of the checks.

"What kind of writing do you use, Pan? I can read some Spanish and a little English, but this isn't written in either of those languages. You better translate them to me yourself."

He handed Pan the packet of checks.

The Top Hand leaned his chair confidently against the wall by the bureau and started to read.

"Check Number 21 to Kluskus Tommy—$16.70 for wages less $6.80 poker debt owing Lester Brewster." Pan stared hard at the check and scratched his head. "Ya just can't beat Lester at wild horse stud," he said.

"And here's check Number 14," Pan read hopefully. "To Ole Nucloe Localo." He had trouble pronouncing the name. He looked up half apologetically at George.

"You know Ole," he said. "Sixty-six dollars for bay horse, also wages, less $12.50 Lester Brewster poker."

George still hadn't written anything down.

"That's a hell of a way to keep books," I put in.

Pan hurried on to the next check.

George interrupted him. "Just a minute, Pan. You can't even figure out what these checks are for. You boys have got to figure them out yourselves in the evenings, and for God's sake you'll have to keep better books and records. This is an incorporated company. The U.S. and Canadian governments as well as the shareholders are going to be on top of you one of these days, and they're plenty strict. One look at a mess like that, and, as I said before, they'll hang the both of you."

CHAPTER II

The Girl

ON OCTOBER 29, 1939, when Pan and I finally drove the heavily laden bunch of pack horses across the narrow Quesnel bridge, high above the Fraser River, and headed for the Frontier, we had no real conception of the hardships and problems that lay ahead of us. If we had, I doubt very much if I would have tackled the proposition.

George Pennoyer had ridden out of Quesnel several days before, planning to gather up all the cull cows, doggie calves and yearlings on the Batnuni Lake Ranch. Upon completion of a satisfactory gather, he and the two remaining Batnuni cowhands were to trail the beef the seventy-five miles on a newly cut trail out to Vanderhoof, a village on the Canadian National Railroad. And thence to Vancouver where he would sell the critters and get enough money to pay our immediate bills. From there Pennoyer would return to his home in Wyoming.

Pan and I rode abreast behind the string of horses. It was not necessary for one of us to ride in the lead as Nimpo, the crooked-footed, blazed-faced, black cayuse who became famous as The Horse That Wouldn't Die had taken his usual position up in front, and he knew as much about keeping a pack train lined out in the direction of home as we did.

Nimpo was a character. He was a hardheaded businesslike individual who not only took care of himself but also looked after Stuyve, his lifelong friend and half-brother. Nimpo's main failing was that he never knew when to quit. When he started out to put over some smart deal that he had been mulling over in his mind, he never gave a thought to his own physical discomfort while doing it. He never turned back once he had started, no matter how hopeless the proposition became.

Nimpo was knotheaded. Nimpo was lug-headed and he was boneheaded. But he knew how to handle a bunch of horses. He was a dandy leader and stood for no nonsense. We moved westward at a steady, ground-gaining, three-mile-an-hour pace.

We caught an occasional glimpse of the old Bear who, as usual, held down his self-appointed scout position some two hundred yards up ahead of Nimpo. The Bear never allowed a man or a horse to take what he obviously considered the most honored and important position. I don't think he considered any of us capable of holding down such a strategic post. It was always amusing to see the serious-looking, long-faced, shaggy-haired Bear strutting proudly down the road, his tail and his head in the air, in the lead of a string of horses or a bunch of cows.

Pan fumbled in his pockets for his package of tailormades. He had no luck. I knew what was coming next.

"I guess you got aholt of my coffin nails," he said. "Pass 'em over."

He stuck his hand across his saddle horn in my direction. It was typical of Pan. He knew I'd get a jolt and wanted to see the expression on my face. He would start this trip off in his usual practical-joking way.

I was riding the Spider. The high-strung roan made a fast, crooked pitch to the side of the road. I didn't say anything. Pan kept his hand out in my direction. I reined the Spider in close and handed the Top Hand a half-filled package of cigarettes.

He tapped one out, stuck it in his mouth, then stuffed the package in his own pocket.

Now the match-and-lighter fumbling started. I watched him.

"Match," he said, and once more reached his hand out in the direction of my saddle horse. Spider jumped to the side of the road again.

I contemptuously handed him over a match. He lit his cigarette. We rode on in silence.

For the first nine miles out of town our pack train passed occasional homesteads and small farms. Once a battered old truck slowed to a stop to let the train slip by. The driver waved and grinned. I said, "Howdy."

Nimpo stared hard at the man for a moment, moved slowly on around the truck and then struck out at a slightly faster pace.

We were now riding into a great chunk of British Columbia's burned-over wilderness. Blackened, charred snags, rotting down-timber, windfalls crisscrossed and piled in ugly profusion, reached away on all sides of us.

Far in the western distance a lush, green-colored mountain rose against the sky to break the desolate monotony caused by some careless white man or Indian with an irritated match or a hungry cigarette butt.

Looking back over my shoulder I caught a last glimpse of the twisting Fraser, the river with the fabulous history which was to be so excitingly described by Bruce Hutchison in his book *The Fraser.*

As Pan and I rode abreast behind the pack train, I decided, after considerable thought on the subject, to break down and tell him about the girl who had been haunting me for the past several months.

Bachelor cowhands, packers and the other horsemen who often go for months at a time in the lonely back lands of this cattle frontier without any contact with women, or any visible signs of entertainment other than what they can cook up among themselves, such as Sunday rodeos or practical-joking tricks like Pan pulls off, often reminisce about their past love affairs. Nine times out of ten, the plot follows the same pattern—the girl that they didn't quite get.

A cowhand, after cautiously approaching the subject, is usually goaded into breaking down and telling all by his riding partner. There is a certain etiquette that is adhered to in such soul-searching moments. The listener stares down at his saddle horn, listening intently, deep in thoughtful meditation. He appears to be carefully weighing all of the facts of the case so that he can sum up his friend's woman troubles with a few words of wisdom and some serious advice. He knows only too well that it will be his turn next to spin some yarn or relate some tragic incident of the past which sent him riding towards his range, broken-hearted, broken financially, and remorseful—"busted, gusted and disgusted," as the cowhands term it.

I won't bore the reader with the details of my own sad love life, except to say that like the rest of the boys in the back-lands, I always came in on the last lap—the loser.

Of course, after Pan and I had arrived on the British Columbia cattle frontier, my time had been absorbed by the projects at hand. Women just hadn't fitted into this picture at all, until suddenly the situation reversed itself, and I was being haunted by a certain blond girl who, mysteriously and without any logical explanation, had entered my life.

I started to rein my saddle horse close to Pan and the Piledriver so as to tell him my fabulous love story, but noticed that three-quarters of an inch of ash clung to Pan's cigarette. This put me off for a moment, for Pan had recently taken up the nasty habit of seeing how long an ash he could grow on a cigarette.

I knew that Pan cultivated his new cigarette-ash stunt to irritate those people who were unfortunate enough to be in his immediate vicinity. The trouble was that Pan's hand was always as steady as a rock, and he had become as adept at this prank as he was with a lariat rope, a six-shooter, or a snaky horse. Usually, when the ash had almost reached his lips and the whole thing had begun to bend in the middle, the people around him would get panicky and frantically tell him to knock the ash off before it fell into his coffee cup or on the new rug. Here was just where Pan fooled you. He would suddenly reach across and the one and one half inches of ash would drop into your pocket, down your neck, or on your nice, new, silk scarf.

But luck was in my favor this time, for Pan's saddle horse suddenly stumbled over a protruding windfall, knocking the ash over Pan's chaps. I saw that the coast was clear. There was no chance of getting cigarette ashes down my neck or in my ear.

Riding over close to the disappointed Hand, I cautiously eased into the problem that had, on numerous occasions, kept me awake nights.

"Pan," I said, "there's something been bothering me. Something I can't help at all—something I have no control over, and this is just as good a time as any to ask your advice on the subject."

Pan swung his gleaming, narrow eyes towards mine, and an amused look came into them.

"Easy there, friend," he said. "Easy—you sure ain't gonna tell me about some woman. I've been around the stockyards and cattle-loading pens with you this trip to town, and I didn't see any big herd of ladies tracking you down—not even one cull."

"It's a girl, Pan—a young girl with blond hair and almond-shaped eyes. She's got a figure that's out of this world. Full lips, shaped like a cupid's. I've seen her lots of times."

The Top Hand scratched his head. "I ain't seen no gal like that around here, friend. You musta been dreaming."

"Not exactly, Pan. I was asleep the first time I saw her all right, but it was different from a dream. When I woke up under the spruce tree, she was still there, smiling at me, and she was whispering something. I couldn't make out what she was saying. But, when I rolled out to throw a stick on the fire, she disappeared."

Pan wet his lips. His eyes had a faraway look. "A dream," he said. "Just a dream girl—how old do you figure she is?"

"All right, let me finish," I snapped at him. "This same girl comes back, over and over again. She's been dressed differently at times. Her golden-colored hair was long at first. It reached almost to her shoulders. But a short time ago, she must have had it cut. Now it's short curls and a small kind of a bang on her forehead. Pan, I just can't figure the whole thing out. I'm crazy about her—I've fallen like a ton of bricks."

The Top Hand was very serious now. He was staring hard at his saddle horn. "You know, friend," he mused, after a long silence, "that there blond girl could sure as hell be alive all right. Maybe an old monk sittin' on top of Mount Everest put her on your trail. You'll probably meet that gal some day. Sure you might. Maybe it will be a long time before you do, but it'll save you lots of woman trouble, son, if you keep her for your girl-friend."

And then, a sad look suddenly swept the Top Hand's weather-scarred face. "And, anyway, friend, it's just as well she's not here in person. These sticks are too hard on a woman. I ought to know. Ya saw what happened to my marriage."

I understand the psychologists have a more scientific explanation than Pan for this sort of thing, though still not very satisfactory to the skeptics. It seems that many men create an ideal female out of their imaginations and then proceed to spend a large part of their lives seeking her—or a reasonably good facsimile. However, I like Pan's notion better: that an old monk on Mount Everest put the dream girl on my trail.

CHAPTER III

Pan Makes a Deal

OUR IMMEDIATE PROBLEM was to line up hay for winter feed for the Batnuni Ranch unit. With the sudden suspension of haying operations on my main unit, the Batnuni Ranch, we had been left in a bad predicament. If we could not line up some eight hundred tons of hay, we would be forced to sell at least five hundred head of top breeding stock. This would just about finish that end of our cattle operation. Now Pan had an idea.

"Friend," he said, "our main hope of lining up hay is from old Joe Spehar, whose layout sits on some big swamp meadows this side of the Nazko Indian village. I figure that unless we can make a deal with Joe, we'll be beat. There's a way of gettin' around old Joe, and I'm tellin' you now this is one deal we can't muff. When we get into Joe's ranch, you just hold your mouth closed, and let me do all the talkin'."

"Yes," I said, "big-talking, big-brained Panhandle Phillips. You do the talking all right, on this deal, but God help you if you make a mess of it, and I have to go in there after you, and straighten it out with old Joe."

Pan snorted like a wild horse, then yelled at Nimpo as we approached the turn-off into Joe Spehar's hay ranch, but he didn't have to urge the smart cayuse. Nimpo knew where the best hay in the country was piled up. He swung in his tracks, bit Big George in the shoulder, then darted around the bewildered horse, turning him and the rest of the horses into the trail. The remaining cayuses now strung out behind the leader. Pan grinned. "That knotheaded old buzzard sure knows where the feed lays, and he can herd a bunch of horses towards it better than most horse wranglers."

The trail cut through the timber in a straight line, dipped down through a mud-bottom creek, where the horses sank almost to their bellies before they struggled out on the opposite bank, and then ended suddenly in a stump-sprinkled, jackpine clearing, where a row of old, weather-grayed, log-sided, shake-roofed buildings sprawled awkwardly around each other in a bent and twisted three-quarter circle. We were at Joe's ranch.

I could see the wide, yellow-colored opening of the huge meadow through a straggly group of trees on the far side of the clearing.

One end of the front porch of a long, log building had caved in, and the upper panel of the battered front door looked as if someone had recently smashed it in with a club or pickaxe. It hung by one hinge, slightly catawampus.

A five-foot-tall man, with a moon-shaped face completely surrounded by a fringe of white beard, and with a bald head ringed by white hair that stuck straight out like porcupine quills, pushed through the doorframe and surveyed us with completely expressionless, pale blue eyes.

"Howdy, Joe," barked Pan. "We're gonna throw down our beds. Where do we turn the horses?"

Joe didn't answer. He was staring hard at me. He'd only seen me once before. That had been at his summer-range dugout on the middle Blackwater, fifty miles from here. The meeting was almost blurred out by a five-gallon keg of Joe's homemade parsnip wine.

Joe was one of those lucky individuals who are never conscious of themselves or the other person. He concentrates so hard on the subject immediately at hand, in his expressionless sort of way, that he is oblivious of all else. Joe always gives the impression that he has the upper hand at any gathering. He gives you that dangling-in-empty-space sensation.

The horses were now waiting for orders, nibbling at the odd grass blades in front of the cabin and between the stumps.

Pan, who knew Joe better than I did, carried on a one-sided conversation while Joe stood, motionless and expressionless, staring at me, apparently not listening to Pan. I didn't say anything but grinned and waved my hand at him in a self-conscious attempt to look nonchalant.

Finally Pan barked at Joe: "How's your hot-water situation? We've got a case of O.P. rum and it's ready to get opened and heated up."

Joe's eyes flickered faintly, then he pointed a finger at my saddle horse and grunted loudly. "I've seen that son-of-a-beetch before."

He turned abruptly and walked into the cabin.

"You're not getting very far," I said in a low voice to Pan.

"Don't fool yourself, friend," he replied. "Joe's putting a bucket of water on the stove."

A moment later Joe reappeared, wearing a tall shako-shaped hat with a long tassel. Now he headed for some half-fallen-in corrals. The pack horses followed him, and Pan and I brought up the rear.

We tied up and unpacked the fifteen horses and turned them, with our saddle horses, into the corral where Joe was dragging out and scattering hay.

When the cayuses were taken care of, Pan pulled a bottle of Demerara rum out of a coal-oil box, flourished it at Joe and was almost run over, first by Joe and then by me, as we headed for the cabin and the hot water at a running walk, the tassel of Joe's shako flying in the breeze.

Joe Spehar is one of this country's most baffling characters. He will give you the shirt off his back if he likes you and believes you to be on the level, but just look out if you cross him the wrong way. Although he is not over five feet in stature, one never thinks of Joe as a small man. He has a deep, rumbling voice, hardly ever smiles, except with his eyes, and can bellow and roar like a bull when the occasion warrants.

Joe Spehar has never been known to fall down on his word. When you shook hands on a deal, that business was closed, and no matter what the other man did, Joe always came through. For this reason he did a lot of thinking before making a decision.

Joe claims he was born in a cave in the mountains of Yugoslavia. He was one of the early arrivals in the Nazko Valley, and buying and selling fur was his specialty, but he kept a roving eye and a sensitive ear on all cattle-ranching possibilities.

Several years before we came into the country Joe bought out a settler who had been starved off a half-section of meadow land on the edge of a three-thousand-acre swamp. Joe ran a practised eye over the contours of the big swamp and figured it could be drained. He purchased and filed claim to strategic quarter-sections scattered across the grassy bog.

Nazko settlers told him he was crazy. The fur market collapsed and Joe lost everything but his spooky-looking swamp land. He made a deal with some Nazko Indians who owed him money to drain the upper end of the big swamp.

It took a lot of sweat and optimism to dig two miles of drain ditches through a mosquito haze so thick and violent that moose left there in the summer months—but Joe stayed with it. The swamp was drained, and a terrific growth of red-top, bluejoint, wild millet and Reed's canary grass sprang to life. This tangle of grass grew higher than a man's head. You could actually get lost on a cloudy day in this jungle of hay.

Joe knew he had something. He traded the Nazko Indians hay for cattle—opened a small trading post on the edge of the great new meadow. Once again he bought the Indians' fur, but this time he also traded them grub and hay for cattle. By 1934 Joe had collected some three hundred head of cattle and paid for nearly all of his land.

But the depression was at its worst up here then, and that year Joe's beef brought only one and a half cents a pound. He had shipped four carloads to Vancouver. They didn't bring enough to pay the freight. The railroad took Joe's beef money and sent him a small bill besides.

The bank took this opportunity of closing down on Joe. They refused to renew his note. He had to pay at once.

Joe talked to the manager. They were not lending any more money on land or cattle. Joe realized that at that point his cattle and land would not bring enough money to pay off his loan. He had bills to pay as well. His biggest debt was to C. D. Hoy, an honest, big-hearted Chinese who was building up a general-store business in Quesnel.

This was the first time in his life that Joe couldn't pay his bills. He cursed the banker in a loud voice. As he opened the door to leave, he roared, "Some day you gonna come to Joe Spehar and ask him to give you his business and his money again!"

Up the street at C. D. Hoy's general store Joe sat down in the back room and relaxed. The Chinese had a great eye for business and he could sense a man in trouble. He had a tiny office, where he often invited his customers. In the file-jammed little room C. D. would pour, in a regal but calculated manner, drinks of the best Scotch whiskey into exquisite Chinese glassware—and this is what he did with Joe.

Joe gruffly explained to C. D. that he was unable to pay his immediate bill.

"You got to wait," rumbled Joe. "Maybe long time— but I pay you interest."

C. D. Hoy listened gravely to Joe's troubles, then asked him how much he needed to pay off the bank and how much more was needed to carry him for another year.

"I don't know what's a matter with most of these banks," said C. D. "They must know that two and a half cent is being knocked off the duty to the States this winter. Cattle are all tied to Chicago market. Then two-cent beef will be four cent by spring. Nobody can lose if he buy beef now."

He looked sharply at Joe.

"I'm going to buy one hundred head good Hereford. But I also buy good registered bulls for total of three hundred head of cows. You run my cattle with your bunch. Use my bulls. Ship my beef with yours. I pay my share cost of hay and wages and grub. I sell you machinery and grub from my store. I lend you money pay back bank today. Pretty soon we both make lots money."

And that was the beginning of one of the most efficient partnerships I know of. Within five years' time Joe had not only made enough money to pay off all lands and debts and drain his huge swamps, but was able to contribute generously to any and all worthy causes in that part of the country. Today Joe has a model ranch capable of feeding fifteen hundred head of cattle. He is comfortably retired and plans to lease the big layout out on shares. And C. D. Hoy made a good thing out of his investment.

That's just how close Joe came to losing his cattle, his land, his self-esteem and his shirt. It took a thrifty, business-minded, visionary Chinese to save the day. Many other ranchers were not as lucky as Joe during that tough period, for there were no Chinese in their neighborhood.

Although C. D. Hoy is one of the most successful businessmen and merchants in this country, there were many more of his countrymen with less worldly goods who came to the rescue of those they trusted. These men (many of whom could barely read or write English) were the ones who, during the tough times, took the place of the banks to bolster up the economy of B.C.'s Cariboo and Chilcotin districts. They are much respected and highly regarded pioneer citizens of this north country.

That night in 1939 at Joe's, Pan sat on a log stool near the stove and did most of the talking, occasionally snorting like a horse to emphasize some point. Old Joe leaned back on his low wooden bunk with his feet stretched out on the hard dirt floor, stared at the floor, and at well-spaced intervals made strange grunting sounds.

Finally Joe cleared his throat thoughtfully, and agreed to winter three hundred head of cows for five and a half dollars apiece if we furnished a man to do the feeding and supplied his food. He also agreed to rustle up to fifty head of our horses for nothing, if our man kept them out of his stackyards.

This was a lucky break and a dandy deal for us. Pan and I rolled contentedly into our sleeping bags on the dirt floor in a far corner of the room.

At daybreak I heard old Joe clumping about the cabin, starting the fire. Suddenly he was standing above Pan and myself, roaring like a mad bull.

"Snow—snow—son-of-a-beetch—she snow. Too early for snow—my cattle steel on range."

Pan was up and out of his bag before I was. He stared out of the window and snorted loudly.

"This changes our plans right now, friend," he said to me. "I've got to cut out my pack horses from yours and hit the south trail for the Home Ranch units pronto, and you've got to move fast for the Batnuni ranches."

Pulling on my pants I looked through a half-broken pane of glass to see nearly a foot of snow on the ground. It was falling so thick the meathouse and barn were obscured.

This was dynamite. I realized our troubles had started. There would have to be some clear thinking and some fast action. Some mighty tough slugging lay ahead of us.

CHAPTER IV

———— ✦ ————

Snow and Poison

AFTER THE FRONTIER COMPANY had been organized, Pennoyer, Pan and I divided it into two main parts. Pan was put in charge of the Home Ranch area. The Home Ranch headquarters was situated on the remote and little-known headwaters of the Blackwater River, the country I described in my book, *Grass Beyond the Mountains*. It was over two hundred and ten miles by pack horse and wagon trail from Quesnel. Around the Home Ranch we had built cabins, barns and corrals on large, wild meadows at the Rich Meadow, Irene Lake Ranch, Tommy Ranch, Alexi Meadow Ranch, Sill Place and Itcha Mountain Camp. These units were all located within forty miles, or a day's saddle-horse ride, from the Home Ranch.

A hundred and ten uninhabited miles east of the Home Ranch, I was cow boss in charge of all operations in the Batnuni area, which included the Pan Meadow Ranch with hay ranches on Bear Creek and Bog Hole, the McNutt place, the Burnt Desert Ranch, the Pennoyer Meadow and the main headquarters at the Comstock Ranch on the Batnuni River, one hundred and thirty miles northwest of Quesnel.

During 1938 and 1939, up until the time when "the men and money had flown away," a large development program had been going on. Gangs of men, mostly Indians, were at work putting up hay, draining meadows, building cabins and barns, corrals and fences, and slashing out trails. More than two hundred horses, saddle and work stock, were in constant use, and we had trailed in foundation herds of fine Hereford cattle from Alberta, the Nazko country and the Chilcotin.

Before noon, Pan and I parted company near the Nazko Indian reserve, some nine miles beyond Joe's meadow.

Pan drove ten head of pack horses, with Piledriver in the lead, due west on the old Indian trail towards Kluskus Indian village and the Home Ranch. He had a hundred and fifty miles to ride, fourteen inches of snow to buck, and below-zero weather to face in summer clothes and light underwear. I didn't envy him. If all went well he would reach the Home Ranch in seven long, cold, trail days.

With the Bear breaking trail up in front and Nimpo in the lead of five pack horses, I swung north through the Nazko Valley on the trail to Batnuni. I had less than seventy miles to go, one small river to cross, and two cabins to stop in, en route. Luckily, I was wearing my long woolen underwear. I hoped to reach Batnuni in three days, since my pack horses weren't as heavily packed as Pan's.

I arrived at Batnuni Ranch on the night of November fifth. Meantime, George Pennoyer and the boys had rounded up and cut out about one hundred and thirty head of beef to make up five carloads. In the morning he and two top cowboys, Rob Striegler from Vanderhoof, and the Chilcotin Indian, John Jimmy John, would start the cattle out on the eight-day, seventy-five-mile drive north to Vanderhoof on the Canadian National Railway.

I'll never forget that November day.

The plan was that I ride fifteen miles to the McNutt place, cow-camp at daybreak, check up on the three hundred head of purebred cows which were grazing there, and ride back the following day to Batnuni Ranch to see that our new man, Harold Baxter, was properly feeding some two hundred weaned calves and doing the various necessary chores about the place.

I was then to break into my routine riding job of covering the upper Batnuni Range where another three hundred head of grade cows and yearlings were rustling on some big, open sidehills.

I was to keep this up for two weeks, when Rob and Jimmy John would be back from the beef drive to Vanderhoof. They were supposed to arrive back with Rob's sixteen-year-old brother, Ed, and another man if they could possibly line one up. We needed this extra help to make both the winter drive of the purebreds to Joe Spehar's, and the other drive of the three hundred cows and yearlings to the Pan Meadow, approximately sixty miles by wagon trail southwest of Batnuni.

The Pan Meadow swamp was a big, mud-bottomed, bog country which we had partly drained, and where that year we had built a bunkhouse, barn, and log-fenced wrangle pasture. We had been able to get up enough hay to winter two hundred head of cattle. The other hundred head were to be wintered by a Kluskus Indian named Sundayman Lashaway at his meadow some eighteen miles by trail in a round-about way from the Pan Meadow.

Because of the sudden exodus of hay hands, the termination of finances for the development of meadows, and the resulting reshuffling of plans, the last cattle drive to the Pan Meadow would have to take place at least two months later than the northern winter usually permitted. We would be driving cattle until well into January, through an uninhabited, little-known country. Deep snow most likely would block the way. Weather could drop to sixty below during this time. There would be night herding and camping out with no shelter to speak of.

The cattle on the long drives would have to pick what little grass the horses pawed out from beneath the snow. (Cows don't paw, they nuzzle down into broken snowdrifts pawed out by the horses.)

We would have the dangerous Blackwater River to cross, either through swift-flowing rapids or over dangerous ice, and it was always questionable whether we would be able to pound unwilling cattle into the current when we got them there.

At 4:30 A.M. on that famous November sixth, three hours before daylight, I shiveringly saddled up, said goodbye to George Pennoyer, wished him good luck and good riding, and trotted down the trail through a foot of fresh snow towards the McNutt place. That morning I again rode the big outlaw roan called the Spider. The long trip to Quesnel had taken part of the usual morning hump out of his back, but a man had to check the roan around and snap into the saddle fast or he would find his foot hung up in the stirrup and the roan away in a dead run.

Riding through the dark, timber-bordered trail toward

McNutt's cow-camp, I listened to the long, melancholy wails and the gurgling sounds of a pack of wolves, floating out of the forests. I wondered if those slinky devils had killed any cows. Then I thought about the many difficulties and the long, cold trails that reached ahead of me for the next two months. I realized only too well that Pan and I had on our own hook taken over this deal. No one had forced it on us, and it was only because of Pan's and my stubborn optimism that the financiers of the company had allowed us to take this chance. We would need plenty of good luck and neither Pan nor I could afford to make even one big mistake.

Now it started to snow again. The air was heavy and had a damp, woolly feeling. Snow sifted down like a slowly descending white blanket. This was bad. Every inch of snow that piled up on top of the foot of snow already on the ground would help seal the fate of Pan's and my risky undertaking.

It was fifteen miles between the Batnuni Ranch headquarters and the McNutt slough meadows. Long before we broke out on the first of the openings I heard cows bawling. Something had gone wrong.

When I rode out onto the meadows I knew at once what was wrong—the grass beneath the snow had been grazed off. The hungry cattle, following in the wake of the loose horses, had slicked the country. These cows would have to be moved at once. .

I rode down an incline towards the edge of the creek.

Suddenly, the Spider shied to one side.

The bodies of two cows lay, puffed and bloated, along the bank. Beyond them a bunch of big cows threw their heads in the air and crashed off into the willows. One of them faltered, started spinning around in a circle, and fell heavily into the snow.

I jumped off the Spider and ran to the dead cows.

They were a horrible sight. Their bodies were bloated so badly that they looked more like hippopotami. I touched the head of the first one I came to. She hadn't been dead long. Her finely marked white face was a sickly gray color. Her tongue hung half out of her mouth, black, swollen. Blood-stained froth still bubbled out onto the snow beneath her. She must have died in great pain.

A few yards away, another staggering, half-bewildered cow had regained her feet. Blood-stained foam belched from her mouth.

I straightened up. All around me now cattle were staggering, falling, getting up—going down again. I was ringed by the critters. They were bunching up along the edge of the creek.

And then I saw it—pale, yellowish white bulbs uprooted from the creek bottom—tall-stalked plants growing out of the water from bulbish roots like turnips.

Not a second could be wasted. The hungry cows were feeding on the worst of all poison weeds—the onionlike bulbs of water hemlock. A quick thaw before the snow had popped the roots to the surface. We could lose most of this purebred herd in the next hour.

I started the roan in and out of the crazed cattle. The Bear sized up the situation and sprang into action. He circled in behind the cows, nipped at their heels and howled and barked in unison with my yells.

But these critters were stubborn, balky, ornery. They were hard to smash out of there. It was too much of a job for one man, one horse and a dog.

We would shove a bunch of cows a few yards from the poison creek, pound back through the snow to another group—get them out of there and swing around to find the first bunch back in the weed again.

And then a miracle happened. I looked up from the saddle to see two riders loping through the snow towards me. I was concentrating so hard I didn't get a chance to see who they were, I just saw their horses—big, short-backed, wide-chested bays.

"Water hemlock," I yelled at them.

The three of us without a single word between us had the bunch of cows out of there and on the trail towards the distant Nazko before any more cows went down.

Then I saw who the riders were: Little Thomas, a Kluskus Indian—the nephew of Sundayman Lashaway—and his young Indian wife.

Little Thomas and his wife had been riding towards Batnuni to find out when our cattle would be delivered to the Pan Meadow and to Lashaway's.

I realized there was just one thing to do now—keep drifting the three hundred cows towards Joe Spehar's ranch where they were to be wintered. There was only Little Thomas, his wife, and myself as the crew. There was only a small chance of the three of us getting them there; but there was no alternative but to try.

A faint haze of yellow light slanted down through the veil of falling snow. Judging from the light I figured it to be nearly eleven o'clock. I would have to lope back to the

Vanderhoof trail, overtake Pennoyer and the beef drive, tell George what had happened and urge both the cowhands, Jimmy John and Rob Striegler, to ride back to Batnuni in the shortest order after they had loaded the beef. I would then have to ride on to Batnuni Ranch, give Harold Baxter a pep talk and his riding orders. He would have to take on my riding on the range as well as his own job, for days to come.

I told my plans to Little Thomas. He and his wife were game to throw in with me. They had just one pack horse. Mrs. Thomas drove him along in front of her with a small bunch of lead cows. She and her husband said they would keep the cattle lined out in the direction of Joe Spehar's, seventy miles away. I would catch up to the drive, with extra horses, sometime during the night.

Little Thomas and his wife would have a tough time driving three hundred head of cattle. Moving that many cattle over a narrow trail through more than a foot of snow is a job for at least five experienced riders, and a horse wrangler, all with several changes of horses.

I swung about and eased the Spider into his long distance, ground-covering lope towards Batnuni Ranch. It was one of those days when everything clicked. I caught up with the trail drive to Vanderhoof and hurriedly described the McNutt place fiasco and the subsequent change of plan to George Pennoyer.

George's rocklike face was a mask. He squinted at me and his level, blue eyes were grave. He listened carefully to my hard breathing and excited talk. Then he cleared his throat and spat at the end of a log sticking out of the snow. The beef cattle with Jimmy John bringing up the drag swung around a bend in the trail towards Vanderhoof.

Pennoyer took off his mitts and laid them on his saddle horn. He pulled his tobacco pouch out of his shirt pocket and with three deft movements a cigarette spiraled smoke from his lips. He blew smoke roughly out of the corner of his mouth. I watched him carefully. I knew that whatever this clear-thinking and long-experienced cattle boss had to say would weigh mighty heavy, and I was half shamefaced at my breathless and excited approach.

"Boy," he said, "this is going to be a tough, short-handed job, but you and Pan have played your cards. There's going to be times when you'd wish your dad was here instead and you was safe back home! But you're in it right up to the hilt now, so there's no use me giving you any advice—'cause you're on your own. You've started out the right way—what you did this morning. Keep it up. Good luck and good riding."

George flipped his cigarette into the snow, waved his hand and, neck-reining his horse about, trotted down the trail into the jackpines towards the distant sound of the beef drive.

CHAPTER V

Men of the Nazko

BACK AT BATNUNI RANCH I found Harold Baxter driving a team and a load of hay into the barnyard. Although Harold's well-to-do family owned a farm as a hobby, and Harold had been around stock long enough to learn the fundamentals, he was far from being a seasoned saddle-horse man. But since coming to work at the Batnuni he had been learning the cowpuncher game remarkably fast.

He was a terrific worker, if given his head. Harold was about twenty-six years old, had worked at all the tough, outdoor jobs. His specialty, before throwing in with the Frontier outfit, had been sawmill operating and logging. Baxter stood just under six feet in height, and was hard as nails.

He had a lean face and a well-shaped head. His greenish eyes were set well apart above prominent cheekbones and his jaw was long and axelike. He was a serious-minded fellow who when roused had a temper. On the other hand, he could let go a loud, happy laugh, and I knew he had lots of guts and was ambitious.

I told Harold our predicament and rehearsed him for his tough assignment. Apart from his full-time ranch job, he would now, as I said, have to take over my riding job as well. Neither of these jobs would fall under the eight-hour-a-day category.

"The biggest thing you've got to remember," I told him, "is to be observant when you're riding through those cows on the range. Study each and every one of them. If there's a swollen bag, a lame foot, a snagged eye, porcupine quills in the nose—anything that needs taking care of—cut them away from the bunch and drive them down here to the ranch corrals where you can doctor them. If you have trouble bringing in one cow alone, then drive her in here with a small bunch.

"The main thing is to keep pushing them on to new feed where those upper Batnuni Range horses are pawing back the snow. Keep the critters right behind the horses. You'll be working and riding eighteen hours a day doing a three-man job. If you make good your wages are going up from thirty to thirty-five dollars a month when I get back from Nazko."

Harold laughed. "Five dollars' raise," he said, "that's a lot of dough in these days, but after all, nothing too good for a cowboy—"

There was a bunch of saddle horses out in the meadow. Harold ran them in and I pulled my saddle from the Spider's back. The Spider was in good shape. He'd barely broken sweat; fifty or more miles of slugging through the snow that day had hardly slowed him down.

We packed two horses with trail grub, and haltered the rest.

I tied the six horses head to tail, picked up the lead horse's halter shank and headed towards Nazko just after dark.

I rode Nimpo and led Black Bear, a short-coupled, deep-chested, eleven-hundred-pound half-Morgan by breeding. The Black Bear was five years old and had taken to the saddle like a duck to water. He was one of those horses you hardly have to break. He had come to hand without bucking, running, kicking or even a show of nervousness. With a hackamore, he learned to neck-rein in short order. I knew I had something in this gentle, easygoing cayuse. Although I didn't actually use the full name Black Bear, I will refer to him by that double name so as not to confuse him with my head trail scout, the Bear.

On this trip I led Stuyve, Steel Gray, Montana, Blue Joe and Big Enough.

As we slid down the trail that night I knew that at least I had the horse power to get the cattle through to Joe's.

The dog Bear took his usual position up in front that night. Fifty miles through the snow that day hadn't whittled him down, but I knew he would be played out by the time we reached Spehar's.

It was a brilliant, moonlit night, almost as light as day. The broken track wound and twisted through timber and across open slopes.

It was about ten that night when I saw the Bear bypass the dead cows on Poison Weed Creek. I followed him. Tracks made by the three hundred cattle showed that Little Thomas and his wife had had little trouble moving them for the first five miles. Two well-packed parallel trails, similar to a car road, struck east through the snow in an almost straight line.

And then, all at once, I saw where the cattle had balked. Cow tracks fanned out through the parky poplar country leaving a messy swath many yards wide. It was as if a convoy of bulldozers running abreast of each other had plowed great furrows through the snow.

Cutting off tracks towards the creek which flowed in the direction we were following, I found twelve head of spooky cows which had somehow broken clear of the main herd. I could see them in the moonlight, smart critters that they were, browsing away shoulder-deep in a patch of wild Reed's canary grass.

I tied up the string of horses I was leading and took in after the cows. There was a short, wild ride and we had the bunch out onto the broken trail. Now Nimpo pushed the running cows hard. They wanted to break for the back trail but with the help of the Bear, who was equal to three cowboys, we kept them headed in the right direction.

When I saw the lead cows' tongues hanging out I swung Nimpo around and we trotted back a mile and picked up the string.

It must have been two o'clock in the morning when a mighty tired bunch of cows, horses, one man and a dog slid down the steep hill to the Titetown Lake crossing, twenty-four miles east of Batnuni Ranch.

Up on the bench at the far side of the ford, a giant windfall fire roared towards the moon. The hot, happy glow of Little Thomas' thawing-out fire meant but one thing. The two Indians had shoved the cattle through to the Dry Lake Meadows. The first part of the drive was over, and, with continuing good luck, we would crawl into Joe Spehar's in five days' time.

———

Earlier, I had lined up Letcher Harrington, the youngest of the three Harringtons, who ran small but independent herds of cattle in the Nazko Valley, to move up to Joe's and feed our cattle.

Letcher was a tall, hawk-nosed, tense fellow of about thirty-five years of age who was stone deaf. He always cupped one ear and leaned intently towards the person doing the talking—or, I should say, hollering.

I always got into a yelling contest whenever I encountered Letcher. He outbellowed and outyelled anyone who was attempting to penetrate his cupped ear.

But Letcher was an old-timer who knew cattle and horses and he was not afraid of the longest hours or the hardest work. He and his young son, who had been out moose hunting, saw the tracks of the drive and met our lined-out trail herd two days out of Joe's and gave us a hand.

That night we threw the bunch into Shorty Harrington's pasture and poured three loads of hay to them. This drive had been a lucky one, but I could see another kind of trouble looming ahead when Letcher told me that Joe Spehar was mad at him and wouldn't let him on the place to feed our cattle this winter.

This was a typical occurrence in the Nazko Valley. The ten or twelve settlers in there took turns getting on the outs with each other. It was a kind of progressive cabin fever.

I knew that Letcher and Joe were more or less on a periodic grievance match against each other, but it had never occurred to me that the boys would pick this time to go on the warpath.

Letcher yelled at me from his chestnut horse: "That

miserable little slave-driver—that bald-headed little caveman—claimed I was talkin' behind his back—claimed I called him a Yugoslav."

I yelled back at Letcher that I would have a talk with Joe and see what could be done.

Letcher turned his chestnut about at Joe's gate and called that he would meet me at the Nazko Indian village the next day.

Joe was glad to see me, in his iron-bound, caveman sort of way. I don't remember a single word passing between us while we forked hay to the hungry cows and opened up several water holes in the creek.

That night Joe threw a slicked hindquarter of a big moose bone onto the dirt floor to the Bear, whom he had taken a fancy to. The Bear, who was as discerning of character and high-toned in his choice of friends, had taken an equal liking to Joe.

Then Joe groaned loudly and yelled: "That got-damned Letcher Harrington!"

He yelled this out again, jumped to his feet, roaring and bellowing.

"That Letcher—he say those eggs I ate at Fred Ruden's wasn't poisoned. I *was* poisoned!" roared Joe. "But Letcher he tell all over that I just ate too many eggs."

"Whoa," I yelled across the table. "Whoa there, Joe."

The Bear had now dragged his huge bone across the room to the door and wanted to get out. I walked to the door and gave it a shove. It flew open on its one hinge, and the Bear stalked haughtily out of the house.

Joe's eyes were almost popping out of his head. The fringe of white hair encircling his bald head was sticking straight up and I could see it bristling. He had stopped hollering and before he got going again I cut in fast.

"You say you ate eggs at Fred Ruden's and that you got sick."

"I was poisoned!" yelled Joe.

"How were the eggs cooked?" I asked.

"Fred fried the sons-of-beetches in pig grease," muttered Joe.

"Well now, Joe—how many of those fried eggs did you eat?"

This seemed to shake Joe badly. He bounded to his feet again and pounded the rough lumber table in front of us. I thought his fist was going to smash right through it.

"I ate fourteen eggs," he yelled. "They musta been poisoned. I got seek."

Joe absolutely refused to have Letcher on his place.

This was a bad situation. The feeding job is an important one, and there weren't many men around in the country who could be depended on practically to live with the cattle night and day, good weather and bad, through Arctic blizzards and all emergencies for seven days a week, hauling nearly five tons of hay a day and opening big water holes for three hundred head of cattle.

The following morning I met Letcher in Nazko Indian village, and he had an idea. He shouted his general plan out to me.

He would build a one-room log cabin just outside of Joe's survey line. Letcher knew where the corner posts were. He also planned to throw up a small barn—just big enough to shelter a team, two saddle horses and a milk cow and calf—and a corral in which to work over any hospital cows. Jimmy Lick, a Nazko Indian friend of Letcher's, had agreed to help him.

I promised to pay Letcher the fabulous sum of sixty dollars a month. Out of this he was to buy his own grub and use his own team, harness and sleighs. Further I promised him four bottles of O.P. rum as a bonus in the spring when the job was over.

I wrote out a check, on brown wrapping paper, to Little Thomas and his wife for their superb job of punching cows, and told them I would land the one hundred head of contract cattle at his Uncle Lashaway's sometime between December fifteenth and January first.

Little Thomas was pleased with the check, thanked me and waved and grinned good-bye when he and Mrs. Thomas rode off.

As I crawled up on the Black Bear and hooked the loop of my lead-horse shank over the horn of my saddle I promised Joe that I'd arrive at his ranch once in every three weeks throughout the winter to check on our cattle.

As I rode off leading my string of horses, Joe roared at me: "I steel don't trust that gotdamned Letcher."

CHAPTER VI

Men for the Frozen Drive

THE SECOND NIGHT OUT of Joe Spehar's I unpacked again on the Dry Lake Meadows.

It was a great place for overnight stops, where several small pothole lakes had a habit of drying up in the late summer. By the time snow covered the ground, slough grass grew four feet high on the mud bottoms of these lakes. Here lush green grass sprouted all winter under the snow. It was a miniature cattle and horse heaven if I ever saw one.

Several wide-branched silver spruce and a few jackpines were conveniently grouped together on top of a bench that overlooked the surrounding meadows. Underneath the spruce it was like a house. The lacy, interwebbing branches of the trees shed rain and snow, and dry spruce needles formed a deep, red, sweet-smelling carpet beneath them.

I made an early camp at Dry Lake, staked out one horse and turned the rest loose with hobbles on, below camp. I had just started the fire and hauled up a lard pail of water for coffee when I heard the whinny of a horse float up from somewhere back on the Nazko trail.

I looked down at my horses. They were munching away in the deep slough grass a hundred yards from camp. I thought I must be hearing things, but the whinny came again, and this time all the horses threw their heads in the air, listening intently. Nimpo let fly with his long-drawn-out, melancholy, donkeylike call.

It must be a loose horse following our tracks or some Indian traveling a trapline. No white men would be crazy enough to be traveling this back country this time of year, I thought.

Out of the gathering dusk I saw a horse and rider and then a pack horse break out of a distant draw. I threw some sticks on the crackling spruce fire, dropped a handful of coffee into the boiling lard pail, and looked up to see that the rider was forking a short-coupled, trim-legged buckskin horse.

I could tell by the way the man sat his horse that he was used to the saddle, but the outstanding eye-compeller of the outfit was the stranger's chaps. They blazed out into the gathering dark like the red embers of a fire, a brilliant red-mohair work of art.

I yelled, "Howdy there—come in—the coffee's hot."

The cowboy rode up the hill, snapped off his horse, dropped his bridle reins and stepped to the fire.

"This sure is a right nice layout," he said, looking about while he pulled off his mitts and held his hands over the blaze.

"It's just like home for man and horse," I answered.

The rider was of medium size. His face was weather-beaten and rugged. Character showed in the straight, determined line of his mouth and his firm jaw.

I took an instant liking to the fellow. He threw out a feeling of strength held in reserve.

I poured coffee into two tin mugs and handed the stranger one.

I didn't ask him who he was, where he was going, or where he came from. That is the code of the frontier. They're a man's own personal business. He has the right to tell you all about himself, his past and what his plans are, if and when he wants to.

This unwritten law applies mostly to the clan of rancher or rider—the horseman group. It is noticeable in any cow country but is not practised particularly by farmers or townspeople. It's a good custom too, for a man on the frontier goes by what he is and what he can do, not by what he was, what he's on the dodge from, or where he did it.

I stuck a spoon in the can of powdered milk, helped myself and passed the can to the stranger. He thanked me, helped himself and reached for the sugar sack. He tested his coffee, took a slow drink and set his cup down on the ground. Then he squatted on his boots.

"Nothing like good old java, thick as molasses, and boiled in a lard pail over a campfire," he said. His eyes lit up for a moment and a faint smile flickered across his previously impassive face.

"It gives a guy a kind of warm feeling when it hits the bottom of that empty belly," I replied, as I poured myself another cup.

The rider looked thoughtfully at his steaming mug.

"You're Rich Hobson, ramrodding the Batnuni end of Frontier Company." He eyed me, no expression on his face. "I can't miss on that one because I've been following your tracks from old Joe's for two days."

"You've caught me with a cold deck," I replied, but the stranger cut in fast.

"I heard you were shorthanded. I'm Albin Simrose. Been riding in the Chilcotin and lately Nazko, but it's not my kind of country. I figured I'd slide up to Batnuni to see if you needed a man."

He pulled a package of tailormades out of his shirt pocket, passed the package to me and struck a match on his chap belt.

This is another big break, I thought. I took Simrose up fast. "You're hired," I said. "Thirty dollars a month and board. Unpack your horse and roll down your bed. We'll hit into Batnuni tomorrow."

It was the nineteenth of December before the Batnuni outfit was ready for the long, cold, sixty-mile drive to the Pan Meadow and Lashaway's.

Three hundred head of cows, steers and heifers from the range were gathered together on the wide, snow-drifted, round-up meadow that stretches out below the Batnuni Ranch buildings. These were the cattle selected for this drive—a trek that could quite possibly be their last.

In a big, square corral adjoining the round-up field, fifty mixed work and saddle horses of every size, age and color kicked and played and fed between piles of scattered hay. These cayuses would play a vital role in the drive. They would be up in the lead of the herd, breaking trail—trampling down the two feet of snow so that the sleighs and cattle could follow. At the noon and night camps the horses would paw away at the snow, opening up the grass beneath it not only for themselves but for the cattle as well.

Some of the horses would be night horses for riding herd on the cattle; others day horse changes for the riders. Still others would later be harnessed and used as teams for feeding cattle and hauling out fence logs and firewood on the Pan Meadow. Among them was Nimpo, and his consistent sidekick, the trim-legged, bay saddler, Stuyve.

Steel Gray, a hard-muscled, eight-year-old horse who was changing from steel gray to gray, as most horses do between eight and ten years of age when they acquire what we call a smooth mouth, stood shoulder to shoulder with young, five-year-old, gray Montana, each horse scratch-biting the others mane.

A short distance away, in a small pasture where an open creek twisted through a bunch of willows, two hundred head of weaned calves licked up the hay that had been forked to them in small scattered bunches. Feeding contentedly among them were some twenty old cows and cripples who would later be turned into what is called the Hospital Yard, where special feed and extra care would be dished out to them. Nineteen registered bulls stood in a close-knotted group by a water hole with their bellies distended, a filled-up, contented look on their curly, white, cud-chewing faces.

This bunch of bulls, calves and crippled cows were to remain at Batnuni, where they would be wintered on some three hundred tons of the best clover and brome grass. They were the ones that were going to get the breaks.

This was the day before the drive, and Batnuni Ranch

was a beehive of activity. A kind of high-tension feeling of excitement and haste filled the air. Near the stockyards, on the hard-packed workyards, a couple of the boys were packing grub and equipment on a heavy-duty sleigh and a lighter jumper sleigh. These rigs were to go with the drive. In the workshop men were repairing harness, saddles, halters and other equipment—sewing buttons on their shirts and pants and doing a rough, male job of washing clothes.

It takes careful planning and plenty of organization to get a winter cattle-drive outfit clicking in this far back country. Not a single item can be overlooked. I had spent days on plans and preparations, was open to any kind of advice and suggestions from the boys. Big chances had to be taken this winter and all thoughts about the subject had to be weighed and decided upon before we pulled from Batnuni.

That evening, Peter Morris, the forty-year-old son of the chief of the Kluskus Indians, rode into the Batnuni with news from Pan at the Home Ranch. Peter had short-cut through the Fawnies and the Nechako ranges, one hundred and ten miles in three days from the Home Ranch. His two saddle horses were leg weary and Peter was all in. His face was gray and more the color of an exhausted white-collar worker than a weather-blackened Kluskus Indian. His thin, long-fingered hand trembled slightly as he handed me a piece of brown wrapping paper with Pan's scrawl across its surface. The paper used as an envelope was torn and scraped, showing the three-day wear and tear in Peter's hind pocket.

Pan's writing was hard to decipher, but as usual the message was terse, short-coupled, and straight to the point. I still have the beaten-up, brown-paper letter. It reads:

Friend—snow done plugged up the Itchas and Algaks and I wernt able to shove through to Anahim to herd in those three men I hired last fall. Three feet snow piled up and only Charlie Forester and George Pinchback here to feed on Rich Meadow, Sill Place and Irene Rich Lake layouts and me here at Home Ranch. Will count on you friend to send up any and all help you can line up in Nazko as I can't leave here to line anyone up. We got to move doggies to those layouts soon or feed will run short here for the Home Ranch bunch.

As usual, Pan was lucky, for Rob Striegler and Jimmy John had picked up several of our old cowhands when they took the beef drive to Vanderhoof. The extra men would ride on to Home Ranch after the Pan Meadow drive to feed cattle for practically no wages. I scrawled a note to Pan and gave it to Peter Morris.

For my tri-weekly saddle- and pack-horse trips to the various units I needed a short-cut trail between Batnuni and Pan Meadow. The inadequate map of the Batnuni-Kluskus area showed the approximate position of the Pan Meadow. It was only about thirty miles as the crow flies due south of Batnuni. Our roundabout horse-shoe-like trail of sixty miles presented a mighty tough and senseless obstacle for horseflesh and, also, for myself this coming winter, when I had to ride the wide circle or triangle—Batnuni to Pan Meadow, to Lashaway's, to Joe Spehar's, and back to Batnuni.

I had once slugged through the rough, almost impassable, burned-over area between Batnuni and Pan Meadow and knew that it was possible to axe out a saddle-horse trail through the mess.

Peter Morris had zigzagged his way through the territory several times and told me that it was still not too late to tackle the job. Zalouie George and George Alec, two Nazko Indian friends of mine from Trout Lake country, and Peter were willing to contract the slashing job for a hundred dollars.

I took Peter up on the deal, and that afternoon just before the mercury started to tumble he had struck west to the Home Ranch to deliver my note to Pan and would then ride towards Trout Lake to pick up his partners and start cutting the trail out from the Blackwater crossing—four miles north of Pan Meadow—through the Burned Mountains to Batnuni.

As nobody else would take the job, I had long been proclaimed the Batnuni cook. Rob Striegler took my place when I was away and was just as cranky and ornery about happy table talk as I was. I shoved pots and pans around the table and looked at the group of hungry men who ate steadily, unworriedly and in almost complete silence.

At the end of the table sat the Chilcotin Indian, dark, clean-cut, well-mannered John Jimmy John, a top cowboy and the best tracker I've ever known. The stoical but humorous Indian, at thirty-five years of age, was a perfect physical specimen. He stood well over six feet in his socks, had noticeably wide shoulders, and about the longest reach I've seen. Jimmy had small feet, narrow hips and well-muscled rider's legs; with his temperament, I've often thought that the right manager and trainer could have built Jimmy John into a great heavyweight boxer.

Jimmy John was the Indian foreman—in charge of all the Indians who worked for us in the summer months. He had been working for the Batnuni outfit for the past two years. Jimmy John was a resourceful hand in any bad emergency. Following the Pan Meadow drive, Jimmy was to carry on to Tatelkuz Lake, between the Fawnies and the Nechako Mountains, where his father-in-law, the venerable old Missue, held sway over a tribe of relations and a gigantic forest empire. Jimmy John would spend the winter with his wife and family, his trap line and his horses, returning to Batnuni and his job early in the spring.

Hunched forward in his seat next to the impressive Chilcotin Indian was small, trimly built Rob Striegler, the straw boss, a hard-working, natural-born cowhand with plenty of guts and the brain to back it up—a good man for the rugged trip that lay ahead of us.

I looked at eighteen-year-old, blond-haired, lanky-built George Aitkens, and thought how different his choice of a way of life was from that of the average college-boy friend of his. George's father was the well-known Colonel Aitkens of Victoria, the Chief Geographer of British Columbia, and one of the founders of what is now called Tweedsmuir Park. Colonel Aitkens had helped us establish the boundary lines of the Frontier Company which was now scattered across an area over a hundred miles long, controlling four million acres.

Although young George had been raised in the city, he had the urge to break away from the conventional and plunge into the daring and the unknown, the line that offered the most excitement and adventure. He had arrived at Nazko Indian village, about two years before, in a pin-striped suit, soft felt hat and oxford shoes, headed for the New Frontier development. The sixty-five-mile trek from Quesnel had badly blistered his feet but it hadn't stopped George. He was shuffling slowly and painfully down the Batnuni trail when I picked him up in my heavily loaded freight wagon. George had a good physique and that rare faculty of not minding belittling himself or being the goat if it gave the rest of the gang a laugh.

Harold Baxter sat next to George Aitkens. He had just about given his all during my two weeks' absence on the Nazko drive. The bitter northern wind had frozen his cheekbones and one hand during one of the long, cold eighteen-hour-shifts on the upper Batnuni Range. The calves had been looked after well, and, all in all, he had done a remarkable job for a man relatively new to this kind of work. I had raised his wages, as I promised him, from thirty to thirty-five dollars a month. Harold was a proud man but this fabulous boost in salary had not gone to his head.

I shifted my eyes to black-haired, half-Indian, half-Scotsman Benny Stobie—a product of his environment—the Chilcotin country, cow camps, bunkhouses. Benny's hobby was poker, a game that he always played and always lost at. But this didn't matter much, since all the happy-go-lucky cowhand wanted out of life was the good easy company of other cowpunchers, his rigging kept up in shape, and a couple of top horses of his own. Benny had plenty of cow savvy and he was trail toughened.

Harold Baxter, George Aitkens and Stobie were all to ride on to the Home Ranch to help Pan after the Pan Meadow drive.

Rob's younger brother, sixteen-year-old Ed Striegler, was a big-framed, high-strung and willing kid with a booming voice. Ed was to be horse wrangler and help Rob drive the fifty loose horses ahead of the herd.

Sitting next to Albin Simrose, whom I had picked for the tough, dangerous and lonely job of Pan Meadow cowboy for the long winter months ahead of us, was a small, finely built, thin-faced cowboy character named Pat Cook, who referred to himself as the Lone Breeze.

Pat has made a business out of always participating in, or having some connection with, the wildest, most hopeless frontier adventures. When the Lone Breeze arrives at your ranch, trail camp or hotel room, carrying his famous luggage, bullhide chaps, bronc spurs and hand-braided hackamore bit over his arm, a fellow knows at once that his project has been selected by Pat as the most colorful adventure taking place on the frontier at that time. The day I saw Pat Cook yarning with the boys on the top log of the corral fence above a bunch of horses, I knew that the Frontier Cattle Company had made the grade.

To have Pat along on a deal meant that no matter how hopeless or disastrous the situation became, Pat's drawling references to some much worse catastrophe would always cheer everyone up.

That evening while the boys did the chores, I checked on equipment and went over winter plans for the last time.

The break we needed most was good weather—but an hour before dark some strange atmospheric pressure or an unnoticeable shift in the wind had started the mercury down, slowly at first, then faster. At two o'clock in the afternoon the thermometer had been steady at about ten above zero. At six o'clock it read ten below. From that point on you could almost see it dropping downwards— until at ten o'clock when we rolled in it had plunged to thirty-five below zero.

Things didn't look too good when I rolled in that night. I tossed around, rolled, tried to sleep, but kept making plans. It didn't take long before the house colded up and then the timbers started to pop and crack. I stoked up the heater—went back to my bedroll-corner in the back room. My high-strung mind, the out-of-control type, flashed back and over our preparations and what lay ahead of us, again and again.

At 3:30 A.M. the alarm went off in an agonizing reminder that the great moment had arrived and I hadn't grabbed off ten minutes' sleep.

People like the Eskimos, who live in terrific cold, have, through generations of experience, learned how to survive by adapting their clothes and living conditions to the country and the climate in which they live. Not so the British Columbia cowboy. For four months of the year I don't think there is a colder, more winter-chilled creature on earth than the inadequately clothed, frostbitten cowhand of the Cariboo and Chilcotin countries in the interior of British Columbia.

Arctic explorers spend a lot of time, money and thought on their winter traveling outfits—caribou and sealskin pants, boots, parkas and overjackets, Primus stoves, caribou sleeping robes, Arctic tents. In order to succeed in their objectives they have had to adopt most of the Eskimo trappings in country where temperatures range from twenty to seventy below zero.

On the cold, windy grass plains and jungles of central B.C. temperatures also drop to seventy below, and, at a few isolated spots, like Redstone in the heart of the cattle country in the Chilcotin, government thermometers have registered even colder. But the low-paid, perennially broke, horse-poor cowhand who faces temperatures that equal the Arctic in intensity, has never adapted his winter trappings to the fierce cold.

Here is the bunkhouse preparation among the boys the day of our frozen drive:

Simrose rolled out. He groped about in the dark, below-zero house for shavings cut the night before. He was shivering. His face was a bluish color beneath his beard stubble. He must have spent a cold, uncomfortable night, but he was too cold to talk about it.

He lit a match to the shavings. The fire crackled. A vague heat began to reach out in a four-foot circumference around the stove. Simrose's bed was on the floor— an antique patchwork quilt, cotton army blankets, and the remains of several old canvas pack mantles. Simrose slipped hurriedly into his Levi pants.

Jimmy John crawled out of his pile of sugar-sack and

old wool-sock quilts made by his wife. He had slept well. Sugar sacks are fairly warm when there are enough of them sewed together.

"Where the hell's my socks?" whined the Lone Breeze from the bed-strewn floor of the back room. "Somebody's got my extra pair of wool socks."

"Quit beefing," cracked Rob Striegler, struggling out of his three homemade wool quilts. "I lost two pairs yesterday."

"Two pairs socks ought to be enough for any man," said Baxter, feeling his sore cheekbones which the frost had left red and peeling.

Soon every man had located his scattered clothes and was ready for his plunge outside.

The boys were all wearing the same clothes they had used when feeding cattle during above-zero weather, with the exception of socks. Everyone dragged on three to four pairs of cheap, light wool ones, then laced up their moose-hide moccasins which in turn fitted into low rubbers with felt inner soles.

This all sounds like adequate footwear. It isn't. There would be some frozen or frosted feet before the end of the drive if the cold snap kept up.

CHAPTER VII

—◦∞◦—

The Frozen Drive

STARS FLICKERED FAINTLY, then vanished in a pale, empty sky; fine frost settled slowly down around us, and a vague whitish steam rose like an eerie curtain from the wide expanse of Batnuni Lake. Far-off sounds seemed to be magnified beyond reality. In the distance, frozen lakes rumbled, then groaned, then exploded like cannon shot into the frosted silence.

Cows bawled and coughed in the dark and the intense cold. The close-knotted silhouette of their dark bodies sprawled out in bold relief against the vast white world around them. The air rattled and crackled, as the cattle called their cold anguish into the dawn of the forty-below-zero day of December 20, 1939.

Out in front of the bunched-up cattle I heard the snap and rattle of hoofs pounding across hard-packed snow; the harsh grunts, the loud breathing of fast-running horses, and then—like the wind—the sounds were dying in the distance, and only a high yell of a rider drifted back to me from far up ahead. The horse cavvy with Rob and Ed Striegler had flashed past the drive into the frozen haze of the early dawn.

Beyond the frost curtain I heard a horse whinny, the

steady creak of sleigh runners, and Stobie calling, "We're lined out boys—let the little doggies roll."

I could hear Baxter talking to his team.

Now the sleighs were ahead of us.

Every unit was in position. All about us cattle shoved and horned each other in the deep snow.

"Wowie!" I yelled. "Take 'em away!"

My job was lead driver. I was to push a small manageable bunch of lead cows down the trail ahead of the main herd. The rest of the cattle would follow these leaders.

I reined Nimpo in behind a group of willing-looking animals on the front end of the herd. With the help of the old Bear, we crashed them out, clear of the rest. The experienced black cayuse and the long-haired, cagey dog filled all gaps. The cattle slugged through the snow towards the road.

I yelled, "Now—let 'em string boys!"

"Wowie—yip!" yelled Jimmy John.

My leaders were clear of the main herd and up on the hard-packed trail. Smart whiteface cows paused with their heads in the air, then swung east towards the pound of hoofs and the dying squeak of sleigh runners. And then they strung out behind the sleighs, single file and in twos and threes.

It was a good drive day. The fine, falling frost soon vanished and the sky turned to emerald green.

By afternoon when I got a chance to leave the leaders in back of Baxter's sleigh, the country had opened up. High bare hills rose from the shoreline of Batnuni Lake. I rode up a long, open slope and reined in Nimpo for a breather.

Here I could see for miles. Far below us, the broken trail twisted and turned above the frozen lake. In the dim distance a small spiderlike body changed shape and color as it moved eastwards across the endless expanse of white. I knew this was the horse remuda, the advance guard of the drive. And then came the chuck and equipment sleighs. They moved slowly and steadily like small beetles across the bottom of a great white bathtub—and behind them moved the cattle—a dark, twisting line nearly two miles long, like a hair snake whose winding body kept breaking in sections and then coming together again.

I watched the distant drive. The Bear sat on his haunches, his tongue hanging out, and looked off into space. My saddle horse stared at the scene impassively. The three of us were watching a rebirth of the old-time, western cattle drives, for this trail herd was moving across the last cattle frontier on our continent. I stepped off on the uphill side of Nimpo and tightened his cinch for the steep descent to the cattle.

That first day was a cold but lucky one. The cattle strung nine miles into camp at our junction corrals with little urging. There I waved good-bye to Pat Cook who reined his horse around and trotted back in the direction of Batnuni where he was to feed the calves and the hospital bunch until our return.

At the junction, two trails split—one running north through the now-impassable Tatuk Mountains to Vanderhoof, and the other southeast towards Dry Lake, Nazko and Quesnel.

We had hauled several loads of hay to the junction

corrals from the ranch. The cattle went hungrily to work on it. Rob and Ed had drifted the remuda on to the McNutt place, one day's cattle drive from the junction, dropped the horses there and retraced their tracks to the corrals.

We strung up two twelve-by-sixteen-foot canvas flies to protect us from any wind that might spring up during the night. I had packed some frozen bannock and frozen stew meat along. These, with rice and coffee, made up our evening meal. The boys were cheerful, the horses in good shape, there had been no mishaps and the tally of the cattle was satisfactory. We were away to a good start.

Several hours before daylight we were frozen out of our beds. The horses were caught and saddled up, the cattle driven down to the lakeshore to drink out of large water holes that had been axed open in the lake. When cattle are eating hay they need water. When they are rustling grass in the snow they can get by without it. The snow they lick up gives them enough moisture to stay thrifty and hold up their appetites.

We broke camp just after daylight and strung into the McNutt place at dark. Sixteen miles and two days of trail now lay behind us, but from this point on we would have a much tougher job moving the cattle. There had been little travel over the trail from here to Dry Lake Meadows, and none from there to Blackwater crossing and the Pan Meadow. The stock would be moving into a strange, new country away from their home grounds.

And now the worst thing that could happen, did. It started to snow. A strong wind swished out of the east, pounding the snow in great gusts and eddies ahead of it.

It was impossible to hold the cattle on the McNutt place another night as the feed was slicked. There was just one thing to do—carry on towards Dry Lake.

That was a frustrating day. My lead cows bunched up on me and refused to face the storm. The big thing in a drive is to keep those leaders moving—the other cattle will follow. We fought the cattle, trying to handle them in separate bunches, with no luck. Finally we jammed them into a wedge-shape formation, plunged our horses into them, and tried to bang them ahead inch by inch.

It was growing dark and we had only made five miles. We had eight to go to make the slough meadows on Dry Lake, our nearest feed. Our mounts were playing out. There was no place to camp—no opening to hold the cattle on. Unless we made Dry Lake this night I knew the drive was over with, and the cows would have little chance of surviving the long winter that stretched ahead of them. Cattle would back-trail on us through the timber, singly, in pairs, and in small bunches all night.

The Bear and I had worked all day from the rear of the wedge to the point, and the dog was all in. Two feet of snow was just too much for him.

Jimmy John rode up alongside of me and yelled above the howling wind, "One of us should ride on to Dry Lake—pick up the other boys and a change of horses. We'll never get 'em there without help."

I told him to slug on through. He disappeared in the swirling blanket of snow.

Simrose, Aitkens and I now had a hard time even holding the cattle. They began to shove us backwards on the trail. It was dark when the boys arrived back from Dry Lake.

Baxter had the small jumper sleigh which had been unloaded at the Dry Lake camp. Jimmy, Stobie, Rob and Ed led fresh horses. We shifted saddles and threw the tired cayuses we'd been riding into the lead of the cattle.

Baxter swung the jumper about and the bunch of us piled into the critters from all sides. I got a few big cows, a long-legged steer and a shrimp-gutted, long-horned, brindle milk cow in behind the jumper and Benny and I held them there. The jumper started forward and the critters followed. Now the big thing was for Stobie, the Bear and me to hold the leaders in close to the jumper and for the other boys to let the main herd string along on the heels of our saddle horses. We plunged on into the night and the storm.

It must have been midnight when we hit camp and the rustling Dry Lake Meadows. In the dark I could see the loose horses pounding away with their hoofs to uncover the grass. The cows struggled wearily out onto the rustling grounds.

We started setting up camp on the low, tree-covered knoll that overlooked the meadows. The wind, carrying fine, biting snow pellets, moaned and wheezed about us. I got the supper going. I melted snow for water in a coal-oil tin and a bucket. Baxter strung up the canvases and weighted them down. Rob and Ed and Aitkens numbly strung up a rope corral and caught three night horses which they staked to lone spruce trees down on the meadows.

We gulped down thawed-out beef mulligan, rice and coffee at about 2:00 A.M.—our first meal since 5:00 A.M.

the day before. We tried to dry out our frosted socks and outer clothing near the windfall fire.

Several of the boys were touched in the feet and fingers with frostbite. A coal-oil pan was set out near the fire and the frosted parts swabbed with the liquid until they painfully came to life again.

The wind and fine snow swished and swirled around the camp blowing the canvases loose. It seemed to be getting colder. None of us remembered having been warm for any length of time this day. I took a chance and told the boys that there was no use night herding since there was now little time left to sleep. We rolled into our beds, all bunched together behind a canvas fly. Everyone left his clothes on.

The fire flashed and flickered against the blast of the wind. Cows bawled into the dark, down on the flats below us. I pulled my head under the covers. My brain spun in circles for a moment. Then blackness reached down and I was oblivious to the pound of the blowing snow and the harsh moans of the wind.

And there she was—the blond vision-girl. But neither the girl nor myself were in this storm. I can't remember what kind of a setting we were in—you know how it is in dreams. But through those few hours of that raging storm the girl and I were in complete peace and silence and warmth, and she was telling me in her soft, mellow voice something about how and where we were to meet again. As I look back, we seemed to be sitting together before a fireplace—I'm not sure now, but that was the impression I got. Our future plans seemed to depend on things I was supposed to do, but I couldn't grab on to them. She was serene—hauntingly beautiful—and transferred to me a great feeling of faith in the future. Faith that I would meet her some place—sometime—and in reality. And then I came to with a start!

I was awakened by a wild fluttering and flapping sound like the sails of a ship in a gale—and down came the flies on top of us. I saw Aitkens automatically rise up in his blankets and feebly start to pull up out of the bed—saw Jimmy John shove him back into the soogans. I heard Jimmy yell at him, "You freeze solid out there, Aitkens. Put your head under the gunny sacks and let the wind blow. Nothing too good for a cowboy!"

I was now wide awake. The vision and the dream still seemed to hover about me, but the change in circumstances was appalling. My underwear, wool shirt and socks were wringing wet—and I was freezing cold.

Ever since Panhandle's advice to me to forget the dream girl, I had practically eliminated her from my life. I had finally been convinced that the whole strange love affair had merely been brought on by lack of feminine contact—quite possibly by some kind of a weird neurosis, or hangover from frustrations of my college days. For nearly two months now, I'd been free and clear of her and had congratulated myself that I had shed the slowly encircling illusions of dreams and fantasies that lead to what we call "getting bushed." Now, here it had happened again. I was back in the same swim, and through no fault of my own. I cursed myself.

Snow and wind still blasted through the stricken camp.

I felt something pawing me. I lifted the corner of my bed. The Bear shoved into it. He was covered with snow, but I let him stay.

A short time later, I had started to doze off again. I heard someone stirring about, and then voices. Two of the boys were getting up. Soon I heard a fire crackling in the wind. Something touched me on the shoulder. I pulled my beagle cap down further over my ears and stuck my head out of the covers. An icy wind slashed at my face.

Simrose was peering down at me. "About four-thirty," he said. "Rob is down getting the horses lined up."

The Bear didn't move. He had wedged himself in behind my back. It was the only part of my body that was warm. I groped about under the covers until I found my extra socks. I sat up, pulled my moosehide coat over my sweater, and got to my feet.

Simrose had a bucket of snow water coming to the boil. He emptied half a pound of coffee into it, and began yelling.

"Coffee-time, you mountain bums! Boots and saddles! The herd's broke over the big mountain!"

There was no movement under the jumbled pile of canvas and beds.

I called into the swish of the wind: "Crawl out of it, you lazy buzzards!"

I stepped over to the rumpled-up fly and started to pull off covers and canvases. There was a lot of groaning and swearing. Jimmy John rose up in his gunny sacks.

"Must be we slept all day," he groaned.

Now the gang began to move.

There's nothing that's harder to face than a blizzardy day, with next to no sleep, with half-wet clothes, and no chance to get warmed up before you start. A man on snowshoes, or a man holding the lines and running beside a team, can get circulation stirring. A man pitching hay or axing down timber can warm up—but, in my opinion, just about the coldest job going is the saddle-horse job of driving cattle.

This is particularly true in a wooded or stick country where trails are narrow and cattle are spooky, because you can't jump off your horse and pound your arms back and forth and stomp your feet up and down. Any kind of unusual movement can stampede the cattle, slow them down, or scatter them in the bush in every direction. Range cattle will spook at the sight of a man on foot just about as fast as they will at one on a bicycle or a child in a kiddy car, and once you've rounded up a spooked or stampeded herd it's an awful tough job to get them traveling in the right direction again.

At Dry Lake the seldom-used trail to the Pan Meadow and the Home Ranch of the Frontier Company, eighty-five miles away, branched west from the Nazko-Quesnel wagon road.

We had the wind in our favor, otherwise the cattle could never have been moved. The snow was belly-deep to a cow and well over the hocks of a horse.

Ahead of us the Poplar Mountains rose gently to the west in a swirling white curtain. On this gradual uphill rise the cattle floundered in the wake of the loose horses and the sleighs.

Baxter's team, pulling the jumper loaded with fifteen hundred pounds, finally came to a floundering standstill at the bottom of a steep hill.

We shoved the herd to Baxter's stalled outfit. Some of the tired cows, with their tongues hanging out purple

into the snow, lay down. The rest stood with their backs
to the storm, heads hanging down.

Albin Simrose bucked his horse around the herd
through the deep snow to my lead bunch and called to me:

"Only one thing to do. I'll ride on ahead and catch
Rob with the horses. We'll rope out a team and I'll lead
them back. It's going to take four horses now to move
that jumper. Harold's got extra harness and collars and
stretchers. Hold the cattle here. I'll be back some time
with the team."

Albin caught up first with Stobie and his four-horse
team. The horses had pulled the load to the height of land
and they and the teamster were taking a breather in a
spruce clump. Stobie, an experienced freighter, had
unhooked the horses, and the little group was working
on a clump of slough grass in a willow thicket. Simrose
crawled stiffly down from his horse and helped Benny
start a fire.

This little slough meadow was nearly four miles
beyond the cattle. The two boys made a quick decision
and a good one.

Our plan had been to throw the cattle into Chinee
Lake pothole this night, but the pothole meadows were
still some five miles away from the spruce clump. The
snow was far deeper here on the mountain than down
where we had come from. The cows would never make
Chinee Lake this night. There was just one thing to do.
Albin would overtake the horse cavvy this side of Chinee
Lake, rope out an extra team and lead them back to
Baxter's stalled sleigh. Rob and Ed would hold the rest
of the horses on the pothole at Chinee Lake until dark

then ride back here to the spruce clump, where Stobie would have set up camp and dragged in the night wood. There would be enough horses, including our mounts, to open up a fair amount of grass for the cattle at the willow thicket.

Albin stomped about in the deep snow, pounding his arms and beating his chest until circulation began to stir again. Then he crawled on his horse and rode.

It was about six o'clock, some two and a half hours after dark, when the bunch of us finally tallied in at the spruce clump and the little meadow.

The snow and wind had stopped, but the fierce cold bit through our inadequate clothing. Trees were cracking with frost and blue steam rose above every moving creature into a pale blue, opaque night.

Despite numbed limbs and aching bodies, every man plunged quickly into the task of getting up as comfortable a camp as possible under the circumstances. We would have to ride herd this forty-below-zero night, two men, two hours at a time.

CHAPTER VIII

―――◆◆◆―――

Ordeal at the Blackwater

AT 4:00 A.M. Simrose and Rob came in from their night-herd shift and shook Stobie and myself awake for our turn around the cattle. They reported that so far nothing had attempted to pull out.

I was so sleepy that the cold didn't seem to affect me any more. I could see the small spruce fire crackling and popping into the brilliantly lit night. Stobie and I dragged to the fire, grabbed up tin cups and drank down several cups of steaming coffee. We rolled extra cigarettes over the fire for the cold ride ahead. A man can't pull off his mitts to roll smokes in such weather while he's holding his bridle reins.

The cattle had scared up some grass during the early part of the evening. Now none of them was attempting to feed, but they were walking around in the hard-packed snow. Their hoofs made a brittle, crunching sound that echoed monotonously into the night.

The stars seemed to be right on top of us. They twinkled and blinked and sparkled. The Milky Way was dwarfed by the aurora borealis, brilliant green and white streaks that flashed back and forth like the rays of giant searchlights across the sky.

As Stobie and I rode around the cattle herd I could feel one of my feet freezing. I eased down off Montana and beat my foot against a spruce tree until a fierce pain came into it. In amongst the trees I slapped and pounded my hands together—then rubbed my face and nose until I could feel them burn. The two-hour night guard seemed like a torture chamber, never any relief, never any letup to it at all.

At six o'clock Stobie and I got the fire going again and the snow melted for coffee. Bannock and beef and coffee for breakfast—and then we were off again into a frozen, crackling horizon.

On Christmas Eve we reached the Chinee Lake pothole. The snow was three feet deep, the air was as still as death, and the fifty-below-zero, knifelike cold stabbed us almost into insanity.

George Aitkens was apparently wide awake. He must have been, and yet as his horse carried him into camp he was calling out as if in a nightmare. Both Baxter's hands were as stiff as boards; and Stobie stood by the fire, waving his arms and yelling weakly at each rider as he rode into camp, "I'll tell Panhandle Phillips. I'll tell Pan all about it. I'll tell him what a nice little camp this was. Wait'll I tell Pan."

I had wrapped a silk scarf tightly about my head and ears. My felt beagle cap with earflaps over the top of the silk scarf gave me little feeling of relief. My head ached and so did my eyes.

The Chinee Lake pothole is a great camping grounds in the spring, summer and fall. On a bench a hundred feet above Chinee Lake, slough grass stands three feet high in a series of small interwebbing basins, and stock can get their bellies full in short order. For several miles around the pothole the country is dotted with small, green, second growth poplar. The lake is alive with rainbow trout, and in the spring and fall the air around Chinee Lake is loaded with ducks and geese.

There are several stories afloat of how Chinee Lake got its name but the one that sounds the most reasonable to me is told by the Kluskus Indians.

Many years before, two Chinese followed the Blackwater watershed from the Fraser River to this lake, panning gold. They made camp on the south side of the lakeshore. According to the Indians, the Chinese had quite a stake in gold nuggets. They always carried this gold in pouches on their persons. One day they left camp with their pouches and pans and never returned. The Indians say that white men had tracked the Chinese down and murdered them for their gold.

There was little firewood around the Chinee Lake country—just a few, small, green poplars and an odd, rotten stump submerged in the deep snow. Everything was iced. A sickly fire, not able to warm a man on two sides at once, gave off enough heat to thaw out a beef mulligan on one side of the pot.

We shivered and stomped weakly around the fire, drinking coffee and batting our arms about. Coal oil was rubbed into frosted parts. Off to the south of us Chinee Lake rumbled and groaned.

It was a terrible night. Everyone was so cold, sleepy and exhausted that the danger point had been reached.

An uncanny sense of fatality hovered about us. All the

boys could feel it. Cattle bawled and tromped around and around on the meadow. Horse bells clanged, making a hard, steely sound, and then—far in the distance—a pack of northern wolves chorused a lonely melody that drifted on the faint night wind—then died away into space. Slowly the sad song rose again, this time high and plaintive, then hung for a moment and fell away into the emptiness of the vast jackpine wilderness that stretched around us for hundreds of miles. It was like a weird nightmare that had us in its grip. We seemed to be transplanted to some long-dead universe.

We stretched canvas tarps under the community bed and over it. I told the boys we'd take a chance and forget the night herding. No man in the outfit had enough backlog of heat under his skin to stand up against this cold without freezing solid.

The Bear was a popular figure. Every man wanted him as a bed partner, but the dignified old boy crawled in next to me, pretending he didn't understand the other boys' happy approaches.

It was a fitful, half-conscious sleep with everybody crowding towards the center of the community bed. Rob Striegler and Harold Baxter would end up on the bottom. Then there would be a general reshuffling and a short, silent period, and then everybody was groaning, thrashing about, shoving, groaning and swearing again.

It was about four o'clock when there was a loud clanging of bells. Hoofs pounded through the snow towards camp. I pulled my head out of the covers into the frosty night air. A dark mass of running bodies came so close to my side of the bed that snow splashed on my face.

The Bear was out of the tarp with a roar, his shaggy body flying through the air at the heels of the running horses. The horse cavvy had pulled down the back trail at a run—and we could hear their bells dying in the distance.

Quickly every man was up and out of the cold bed. Jimmy John yelled:

"Cowboy life is an easy life and a happy one!"

Rob and Ed didn't stop for coffee or talk. They ran for their staked night horses and were off before I had the fire started.

The camp thermometer read fifty-five below. Now we'd have to watch frosted lungs on men and horses.

"Gunny sacks for the horses' noses," coughed Simrose.

He started slicing and slashing our supply of sacks into nose-pieces for teams and saddle horses. These sacks fitted over the bridles and dangled loosely below the horses' nostrils. It doesn't take much heaving, running, or pulling in weather under fifty below to frost the lungs of a horse.

Just before daylight I caught the faint sounds of horse bells drifting in from the back trail. The cavvy had been turned and Rob and Ed were bringing them in.

This morning I noticed that the boys ate very little. Our long, sleepless, seemingly endless exposure to the frost was beginning to take its toll. We were all sick from the effects.

Aitkens' eyes kept closing and he was talking steadily, a kind of rambling discourse on nothing in particular. Baxter's hands had to be thawed out again. They must have been frightfully painful, but he just laughed and made his usual statement that it seemed to be getting colder all the time. It was. Stobie's eyes were giving him great pain. They were bloodshot. We rubbed charcoal around and over his eyelids.

"Can't see very good," said Stobie. "Everything's kind of blurred."

Simrose had frosted one side of his face. One of his lower teeth on that side began giving him hell. Before the morning start I noticed that his face and nose were white and marblelike again. I rubbed them hard until a faint glow showed through the white.

"I never had trouble with my teeth before," said Simrose. "Guess it must be the frost."

Rob Striegler had difficulty getting off his horse and hobbled to the fire, dragging one leg.

"Don't think it's just froze," said Rob. "Must have just pulled a tendon, when I crawled my cayuse."

As usual, my feet had what I call "blowed-up" on me. Several toes and the bottoms were constantly tightening up. I had frosted them on trips in the past, and now they weren't able to take it.

Jimmy had an extra pair of giant moosehide moccasins in his war bag. They were four sizes too large even for me. He stuffed these gunboats full of slough grass and insisted that I take off my rubbers and stick my feet into these small haystacks.

"They're too big to fit in your stirrups," said Jimmy. "You'll have to ride without your stirrups, but they'll save your feet. Suppose you don't use them today—then by tonight your feet he's froze stiff. He'll turn black—then somebody, maybe Stobie, maybe me, he'll have to cut 'em off feet. Both of 'em.

"Now I'll show you something," said Jimmy. "Hold up your foot."

I obliged and Jimmy John poured melted snow water over the moccasins. A film of ice formed on them immediately. I was amazed at how my feet warmed up in my insulated moccasins. They were soon the only warm part of my anatomy.

From the clear pre-dawn sky above us I figured it would be a fiercely cold but bright day. To offset any chance of snow blindness all of us now rubbed charcoal under our eyes. We were a rough-bearded, dirty, black-eyed-looking bunch.

Rob and Ed got away with the cavvy shortly after daylight and soon we had the cattle strung along on their tracks. The horses plunged through crusted drifts along the edge of Chinee Lake. Cattle bucked and reared through the deep channel made by the loose horses ahead of us.

I looked back often from my position with the lead cattle. It was a strange sight that greeted the eye. Every single animal moved through a bluish white haze of its own. This steam rose straight into the air for a hundred feet or more. The long, winding line of cows produced a high, twisting mist. Thin as a knife-edge, it zigzagged across the sky above them.

Several times during the morning I saw fine lines of mist rising vertically out of the swamps and the timber far across Chinee Lake. Jimmy John told me the steam sprouted from moose standing motionless in the brush. Once, far up ahead, I saw a rectangular block of mist standing almost still against the pale icy sky. I looked away for a moment and, when I gazed again in the direction of the misty rectangle, it had changed almost imperceptibly into the shape of an egg. Then all at once it was gone. I knew that I had been looking at the haze above the horse cavvy miles ahead of the drive.

The cattle were now resigned to the trail. Occasionally I slipped awkwardly off Montana and led him for a few paces behind the leaders. But this meant floundering around in the unevenly broken trail and almost spooked the cattle. I couldn't warm up much this way and it was such a hard job pulling myself back up on the long-legged gray without the use of my stirrups that I gave up this attempt at keeping the circulation going and stayed fast in the saddle.

This Christmas of 1939 seemed like a day without an end. Far to the south of the drive the sun made its brief appearance above the snowy Chinee Lake Mountains. The panorama of low open hills, flats and swamps that reached around us on all sides changed to a dazzling white world—a breath-taking flash of brilliance that seared but didn't burn. The unbroken plain of white was so strong that it made me squint, and yet there was no warmth from it.

The sun hung above the mountains on its short noon-day arc, then plunged behind a bald peak, and Chinee Lake was left in an ominous, slowly darkening gloom.

The air bit through my clothing. The icy chill that had settled along my spine was gradually being replaced by a numb feeling. I'd start to fall asleep, then wake with a start.

The mercury was still creeping downwards. I knew that as night fell it was approaching sixty below zero. Our own chances, as well as those of the cattle, of surviving this night were running out on us. I thought: We could drop the cattle now and, as long as we were able to cross the Blackwater safely, we could make it to Pan Meadow and save our own hides.

But I shoved the thought out of my mind. We would stay with the cattle. We would never quit them.

As dark began to settle down around us I felt the fear of death gripping me. It was a strange sort of feeling. Then I worried about going suddenly crazy in the saddle—riding wildly in among the cattle, screaming and bellowing and waving my arms. The urge began to take hold of me—I tried to think clearly.

My brain must be getting frosted, I reasoned.

I pounded my beagle cap with my mittens. I kept punching and slapping the back of my head, my neck and my forehead. Gradually that strange feeling left me, and I was able to concentrate on what might lie ahead of us this frozen night—and on our chances of getting the cattle across the roaring Blackwater River.

As that cruel Christmas night deepened, I thought of other Christmas nights: of bright lights and warmth and laughter. Of cheerful greetings. Of music and sparkling jewelry and crystal chandeliers. Of women's soft voices. My mind drifted across the frozen snow-bound ranges towards my mysterious, blond girl—and I thought how I might never see her again.

I thought of homey places, crackling Christmas fires and gay people. Of children calling happily. I thought of another world, far away, of song and light and warmth and life. And then slowly—like the curtain on a stage—the black, silent night dropped around me.

A cow bawled somewhere in the snow behind, and far in the timbered distance the strange sad song of the wolf pack hovered for a moment, then sank into the silence of frozen forests.

Now a heavy, spruce-and-pine jungle stretched some four miles to the Blackwater crossing, where the trail crossed the river, zigzagged upwards for another four miles, climbing nearly a thousand feet in elevation above the Blackwater River to reach the Pan Meadow haystacks and cabin.

The fourteen-mile drive from the Chinee Lake pot-hole to the Pan Meadow was a long one under the best of circumstances, but with short days, and tired cattle fighting terrific cold and deep snow, it seemed almost impossible of accomplishment. But this was the only hope we had to save our cattle. There were no meadows here on the north side of the Blackwater between Chinee Lake and the crossing. There was no place for the cattle to pick for grass, or any openings large enough to hold them together and out of the timber.

If the cattle sneaked away from us before we crossed the river they would fan out north through the jungle and along the back trail for Batnuni. It would be an almost hopeless job to round them all up in the timber and herd them back to the crossing again—and I knew that few of them would ever make Batnuni.

On the start of the trip I had planned to push the cat-tle across the river, today, to a small opening below the Pan Meadow. I had figured that, with an early start and lots of good luck, we would have a fighting chance of crossing by dark. But I had not planned on the terrific cold, this deep snow, and played-out horses, men and cattle.

It seemed to me that the drive had plugged along in the dark for hours when I first heard the roar of the Blackwater rapids. It was a spine-tingling sound that carried with it the threat of disaster. I could feel my pulse quicken. The noise of the river deepened as we approached.

The tired lead cows suddenly stiffened and threw up their heads. This was a bad sign. I could vaguely make out the old, long-horned, brindle milk cow in the darkened trail ahead of me. She had stopped abruptly in her tracks, listening intently. Then she swung about and savagely flung her head and horns into a yearling immediately behind her. He plunged awkwardly to the side to avoid her attack. The frightened cow rushed into the bunched-up leaders.

This was dynamite. Cattle bawled along the back trail. I quirted Montana in at the horn-slashing cow. He crashed against the point of her shoulder. She almost went down—then skirted around me into the cows behind. I swung Montana about and beat the cow out as she plowed down the back trail scattering the cattle into the bush.

A sharp jackpine-stick missed my eye, took part of my eyebrow with it, and glanced off into space.

We got the crazy milk cow turned about in the trail. I figured Jimmy John was a half-mile behind me. Other cows and steers had turned about and were bawlingly contemplating a break down the back trail.

I was able to hold the line for a long minute. Then, using the same unexcited tone that my leaders were used to, I talked them into turning around and heading for the crossing. It was a close shave. If the front bunch had made a real break for it and beat me out I don't want to think of what would have happened to the rest.

The Blackwater River was about three hundred feet wide at winter water levels, at the crossing. It plummeted at a dangerous speed over boulders and rocks, then plunged into a narrow, slow-moving, undertow lake. The crossing seldom freezes over until late in February.

Paul Krestenuk, a powerful, Russian-born, Indian trader, had, with the help of the Kluskus Indians, blazed and sunk sharp poles at intervals in the rushing water to serve as markers. If the water reached certain levels on these poles you knew whether the river could be crossed by teams and wagons, or only with saddle horses, or not at all. It was almost certain suicide to attempt the river if Paul's markers showed it to be too deep for saddle horses, for if a horse once lost his footing on the slippery rock bottom, and fell in the rapids, the current was so swift that the unfortunate animal seldom could struggle to his feet again. A number of Indians and white men have been drowned in the undertow of these still pools and roaring rapids. One of the victims was young Lashaway, a cousin of Jimmy John's, who had drowned three years before when his horse lost its footing and went down.

A few yards above the crossing, water sprayed and shot into the air as it crashed over giant, half-submerged boulders. I wondered how Stobie and Baxter had made it with their teams and sleighs, and Rob and Ed with the fifty loose horses.

The lead cattle broke from the timber out onto a wide chunk of snow-covered ice, and then stood solidly in

their tracks, staring in wild-eyed panic at the black water
that churned and frothed almost at their feet.

I rode hurriedly out to Krestenuk's markers. Montana
cautiously approached the open river. The saddle-horse
marker read passable. I looked off into the rushing channel. Spray was freezing in the air. Protruding boulders
had turned to grotesque ice-topped figures.

The ice flats came to an abrupt end in nearly three
feet of water. A big chunk of rotten ice had broken away
from the bank and I could see signs of scraping hoofs and
sleigh runners. The cavvy and sleighs had obviously
navigated the crossing.

But I realized that our chances of pounding cattle
into the channel were very slim.

Cows were coming out on the sheet ice—bunching
up—shoving and horning each other. I heard Jimmy
John yell. I rode Montana through the back edge of the
cattle and up alongside his dark form.

"Get these leaders into the water quick," snapped
Jimmy, "before they've made up their minds!"

We rode in—a man and horse on each side of the
stalled cows, our mounts pushing against the frightened
animals.

For perhaps five minutes we talked trail talk, trying
to persuade the cattle to make the cold plunge into the
open water. But these tactics were a hopeless failure.
Now there was only one thing to do—hit them hard—
suddenly surprise them—get the first bunch stampeded into the river before they realized what had
happened.

Aitkens rode up.

I had my long-barreled, 38.40 six-shooter strapped tight on the swell of my saddle. A couple of shots in the air above their heads would get a fast forward move out of the herd, but I knew the shots would spook the cattle behind us who were gradually drifting down the trail. That idea was out.

There was no use waiting for Simrose. He was perhaps a half-mile down the back trail moving the weaker animals in the drag. I called to Aitkens and Jimmy John:

"We're gonna hit them hard, boys. All the noise we can make—the wolf howls all together."

Slowly we drifted our horses back some twenty-five paces from the cattle which stood as if hypnotized.

"Give 'em hell!" I yelled.

We charged at full run into the hesitating herd with wild yells.

The river was perhaps thirty feet beyond the front line of critters. Some of them moved forward a few startled steps, then turned abruptly and started to mill.

This effort had failed. Now the bunch had swung about and were trying to get back off the ice into the timber.

"Hold them!" Jimmy yelled. "For God's sake, hold 'em out of the timber!"

We had to ride hard on the flanks of the cows to get them back onto the ice flats. Back and forth—in and out—forward—turn about—pivot here—plunge there. Gradually we worked the bunch into a great square body a few yards out from the timber.

Now I had to think fast. The sudden fight with the cows had warmed us vaguely, but the horses were weaker, and I knew the cattle had balked for keeps. We couldn't hold the herd on the icecap for long.

Simrose rode up out of the dark. He was mounted on Big Enough. The little, chunky bay was just what his name implied. He was as quick as a flash and plenty big enough. He shoved the last of the drag into the herd. Simrose knew at once what had happened. There was no time to waste.

"Rich," he called, "splash on to Pan Meadow. Get the other boys on fresh horses and bring us back a change ourselves. These cayuses can't take much more hard riding. We'll try and hold 'em here till you-all get back."

"O.K.," I yelled.

I swung the leg-weary Montana around the cattle and after some smooth talking and some gentle quirting he slid into the ice water. Spray hit my face. It felt warm. Montana didn't make that deep, hollow gasp that horses make when they drop suddenly into cold water. The icy river was warmer than the air.

The big gray felt his way cautiously out into the channel. I heard the Bear whimper and then saw him trotting along the ice upstream.

The water deepened to the saddle skirts. I quickly swung my feet forward onto Montana's neck. I held my breath. I knew that if my legs got dunked in the swim in this temperature, it would be all over but the shouting.

We were in the middle of the rapids now. Montana stumbled on a rock, started down to his knees. I saw the fast water coming up at me. But the cagey horse plunged forward again, almost went down, and then got his feet under him. We had been carried some distance below the crossing—a bit too far. The thin ice of the lake below the rapids with the river hurtling under it looked close enough to touch.

The timbered shore was coming closer. Montana broke through the ice on the bank. I took a deep frost-biting breath and dragged Montana around to look for the Bear. We didn't have to wait long. The loud-breathing cow-dog swept downstream to land at almost the exact spot we had. The smart old boy had made it across that river before and had learned from past experiences to enter the water far upstream in order to come out at the landing.

It was a long, uphill pull now. I trotted, then walked Montana. The black jackpines engulfed us, towered high into the glittering sky.

The dense forest was silent, forbidding. Occasionally a tree cracked in the distance. Far off and distinct now from the roar of the river, I could hear the pitiful bawls of the cattle.

Less than half an hour later Montana floundered out onto the great opening of the Pan Meadow. I caught a faint glimmer of light on the night horizon. In the distance a horse whinnied. Montana and I and the Bear had made the Pan Meadow, but three sick, half-frozen men, their horses, and three hundred cattle staggered deject-edly on the ice on the wrong side of the Blackwater River.

The dark bodies of horses churned and stomped about in piles of hay in the corral. Montana struggled up the slight incline towards the cabin light. I looked down at him. He was a frost-covered, ice-caked figure of a horse. I slipped off his back, staggered when I hit the snow, fell, then got to my feet and shoved through the door. A blast of heat struck me in the face. I almost fell again.

A single coal-oil lamp burned on a log table at the far corner of the room. A cast-iron heater roared out warmth. Buckets full of snow were thawing out on its surface.

Rob dragged about the cabin. Ed and Baxter lay half asleep on the floor. Stobie had a wet handkerchief over his eyes. He was propped against a bedroll in the corner.

Everyone looked up surprised.

"Cattle across O.K.?" asked Rob happily.

"No," I barked. "Cattle stalled on ice on far side. Can't move them. Get fresh horses for yourselves and changes for us. We've got to move fast."

"Jehovah's Witnesses!" gasped Baxter. He rose up to a sitting position, and staggered to his feet.

Now the boys were throwing on their sweaters and coats. They shoved out into the dark. Rob carried a lantern. I yelled after them:

"What the hell's the matter with you guys, sitting here in a warm cabin doing nothing!"

The door closed. I dipped a cup into the half-melted snow and took a long gulp. I had trouble rolling a smoke. I inhaled a few times—rubbed my hands together over the heater.

I felt lightheaded. Things were far away. The light was dim. I lay down on Baxter's bedroll on the floor. It seemed only a moment. Then someone had me by the shoulder and was shaking me. I opened my eyes. It was Rob.

"Get up and out, boy. Changed your riggin' to Stuyve. What the hell's the matter with you, laying here in a warm cabin doin' nothing?"

The Bear tried to follow me out the door. I knew he shouldn't swim the crossing again. He stood at the door— his head down. He didn't wag his tail. I told him we'd be back—then closed the door on him.

Again we rode into the frozen world—the silent darkness, the never-ending pain of the raw cold. Aching bodies—strange, ringing sounds in the ears. God, how sleepy we were!

Rob and Ed and Stobie, leading a horse apiece, rode ahead. Baxter and I rode abreast behind them. I looked across at Harold. He was holding his bridle reins in his right mitt.

"Riding dude style," I said. "Right hand on reins, old chap, there's nothing like it."

Baxter's voice was hoarse. He didn't seem very steady in the saddle.

"Ha, ha," he tried to laugh. "I guess the joke's on me. My left hand's stiffened up again. Shouldn't have thawed her out so quick. Ha, ha. Guess it's getting colder all the time."

Stobie kept bumping into trees. I told him to let me lead the horse.

"I'm O.K.," he gritted out. "Kind of headache, throws me off balance. But I'm getting better."

Finally he dropped in behind us, but he wouldn't give up the horse he had started out to lead.

The roar of the river grew louder, drowning out the faraway bawls of the cattle. Now we dipped down the last of the inclines onto the flats and moved cautiously out on the icecap. Baxter rode ahead letting his horse pick its way. Rob and Ed and Stobie followed, leading the extra horses. Stuyve and I brought up the rear.

CHAPTER IX

The Great Spirit Rides Herd

THE CATTLE WERE STILL THERE on the ice. Simrose, Jimmy John and Aitkens had held them. How those three men, on played-out horses, had been able to keep that herd together and out of the timber for so long a time remains a mystery to this day, both to myself and to the boys who held them there.

Our reinforcements, men and horses, managed the crossing without mishap and it was a comfort to me to have company this time in the cold and plunging water.

"We've got 'em," croaked Simrose, as we rode in close. "Don't think a one's got back on us."

Jimmy John yelled, "Give me a fresh cayuse, turn us loose and let her buck. Nothing too good for a cowboy."

Baxter laughed again.

Now while the three men were changing mounts, the rest of us fought the cattle back onto the ice.

While I rode Stuyve in and out of the retreating cows along my side of the milling bunch, shoving and fighting them back into the maelstrom, I thought of a new plan—and my responsibility. Silently I made a prayer. I realized only too well that upon the decision that I made now rested the survival of this herd of cattle and possibly the lives of some of the men.

Simrose, his frozen jaw shining white in the starlit night, rode up on his fresh mount. The other boys were behind him. Now I let fly:

"I'm going to pick the way—break trail into the west. Simrose, you cut out a few leaders, every man behind Simrose grab out thirty or forty critters. Jimmy John bring up the drag. Every man hold what he's got.

"We're driving blind up the river into new country. We'll hold everything close together—one bunch behind the other—and we'll keep them moving parallel to the river. It's new country. There are no roads or trails, but we might find a meadow. It's the only chance of saving the cattle."

I took the lead. Simrose's bunch was close behind me. Looking over my shoulder I saw a dark mass of moving cattle spread out behind him.

I rode very slowly through the deep snow, about a mile an hour. Every little while I yelled back to Simrose. He called back to Baxter who in turn called to the man behind the next bunch.

None of us knew the country through which we traveled. Anything might lie ahead of us. Occasionally I heard cows bawling and men yelling into the black forest.

The timber hemmed us in on all sides. Glancing up through scattered openings in the treetops I was able to locate the North Star. I kept a direction almost due west.

The long hours of this night dragged slowly by—each one an eternity. The wolf pack moaned and wailed. They were much closer—close enough to pick up the scent of live beef.

Now some power greater than mind or physique must have taken over. We had all passed the limit of endurance, but a strength beyond our comprehension led us on and on into the night and the unknown. ·

And then—all at once—Stuyve stepped out onto an opening. He plugged a few paces ahead and then started to paw. He lowered his head. The strongest cattle shoved in close behind us. I urged Stuyve further out into the opening.

The land sloped basinlike towards a center. To my amazement I could see slough grass and then fluffy heads of wild millet and redtop sticking clear of the snow.

"Big grass!" I yelled.

Cows crowded up, their bodies opening up the grass behind them. More cattle rushed towards the oasis in the snow.

We had come upon a small, grassy basin. It was similar to Dry Lake Meadows, a chunk of land that was flooded all summer, where grass stayed green under the snow all winter. Until they got their bellies filled the cattle would never pull out of this meadow. I rode over to the boys.

"For the moment," mused Rob, "we've got her beat. Now, if these cayuses can just get us back across that river without dunking any of us!"

"I wonder if the fires are out up at the cabin," said Ed. He was humped over in his saddle.

Jimmy John took the lead. We followed him back through the forest. I remember a buzzing sound in my head, a deep, nagging ache along my back. I kept falling asleep in the saddle—and coming to with a start when my chap legs hit a tree or a jackpine branch smashed me over the head.

I was only vaguely conscious of Stuyve's movements. I lost all track of time. I remember a couple of the boys shoving Aitkens back in his saddle—Stobie speaking of the darkness, saying he couldn't see a thing, his horse was following Jimmy John's horse nose to tail. And finally I remember hearing the roar of the rapids, and seeing gray daylight creeping up on us—the wet spray of river water freezing on my face.

I remember the warmth of the Pan Meadow cabin—my sudden collapse on the bedroll in the corner.

I remember muttering to the boys that we would have to pull ourselves up and out of the cabin within three hours and get back to the cattle before the wolves stampeded them into the bush.

I can still see Jimmy John warming his hands over the heater, the corners of his straight mouth wrinkling up with an amused expression, and hear the big Chilcotin Indian's reassuring words.

"Sleep, cowboys, sleep. Nobody worry about those cows. He's gonna be O.K. . . ."

I awoke with a start, couldn't figure out where I was. It was pitch dark. The Bear was licking me in the face and whining. I raised in my blankets. My head was spinning round and round. The cabin was cold. I yelled into the darkened room:

"Wake up you guys! We've slept all day. My God, the cattle! The wolves!"

The room exploded into action. Somebody lit the coal-oil lamp, somebody else the heater; and then everyone was running and crashing about the room, falling over each other in confusion.

"We've got to get back to those cows just as fast as our cayuses can get us there," I barked.

Rob was looking for his pants. His face was a pale gray. His trail-worn underwear looked almost black in the lamplight. He barked back at me:

"It's darker than the hubs of hell. If they've pulled out of that pothole or the wolves have spooked 'em, we'll never get them all rounded up again."

"Easy there, boys, don't get panicky," drawled Simrose. He was pouring coffee into a row of tin mugs on the table. "We're not beat yet. Where's Jimmy John?"

We looked from one to the other. The Indian was nowhere in sight. A quick examination of Jimmy's bedroll and belongings disclosed that his 30.30 rifle and a box of shells were missing—and on the kitchen table were bannock crumbs. A cloth sack of jerky had been opened and several strips of dry moosemeat were missing.

Now I knew why Jimmy had said, "Nobody worry about those cows. He's gonna be O.K." The evidence was here. Sometime during the day, probably hours before, Jimmy had quietly grabbed up his rifle and a lunch, saddled up, and ridden off to herd the cattle by himself.

It was while I was throwing together a bannock, and the boys were feeding the horses, that a plan to get the cattle across the river began taking shape in my mind. When Simrose pushed through the door, with a cloud of blue steam in his wake, I told him about it. By the time the boys returned from the horse barn, Simrose and I had worked out the details.

Our plan was a simple one—but we were all in such a frostbitten, leg-weary condition that it took two days and nights to get everything ready for the plunge.

Jimmy John had had little trouble riding herd on the cattle. The wolves had picked up their scent all right, but, strangely enough, had left the herd alone. When Jimmy arrived for his lonely shift he found that they had hamstrung, torn to pieces, and gorged up a bull moose at the edge of the timber and within two hundred yards of the cattle.

Rob and I relieved Jimmy John just before daylight. I sent him back to the Pan Meadow to thaw out. Before he rode off he said, "That moose he try to hide in those cattle. Cows stand still and watch wolf—but moose he scare bad and run away. Then wolf bring him down and eat 'em up quick."

We held the cattle on the pothole for two days, with two of us riding twelve-hour shifts around them. A big, windfall fire burned day and night on the edge of the meadow where the back trail took off, and a steadily replenished lard pail of boiling coffee threw its soul-satisfying fragrance into the icy air.

Back at the crossing, the steely ring of axes, the rasping scrape of crosscut saws, the thunder of wide-branched spruce crashing across the snow, the rattle of chains and the thuds of horses' hoofs on the brittle ice, the loud commands of teamsters, broke the monotonous roar of the Blackwater torrent.

Gradually a long, rectangular brush corral reached from the timber across the icecap and into the river. When the teams had dragged the last trees into place, I looked at the eight-foot high, crash-proof enclosure that narrowed down to a long funnel-like snout at the edge of the rushing water.

"Somebody come along here," commented Jimmy John, "and see cowboys building big pasture fence on ice. He's gonna say, 'Those boys gone cold crazy.'"

"Don't worry about that," spoke up Rob. "The ice will be out of the river long before the next damn fools come along."

On the twenty-ninth of December, ten days after the drive strung out of Batnuni Ranch, ten frozen days and nights that had blacked out all other cattle drives in our memories, we shoved the gaunt herd of purebred cows into the great brush corral, and took our positions among and around them.

We let the cattle stand in the enclosure for a few minutes to stare, with drooling mouths and upturned noses, through the narrow mouth of the funnel at Benny Stobie, pitching a sleighload of green, sweet-smelling hay on the far side of the river. When the scent of the hay had been picked up and located by the cows that jammed the funnel opening, I nodded to Jimmy John, Rob and Simrose, who held their nervous horses beside mine, directly behind the fifteen or twenty lead cows. I turned in my saddle and waved an arm at Baxter, Ed and Aitkens, who sat their horses at the back end of the cattle, nearest the timber.

And then all hell broke loose. Baxter emptied my

six-shooter in the air. Aitkens and Ed started beating on two coal-oil tins with small iron pipes, screaming like banshees. The four of us behind the lead cows let fly with a wild chorus and plowed into the already-stampeding cattle in front of us.

These cows jumped, plunged and reared out into the river. And now we were in the channel, the lead cows, with only their heads above the surface, were swimming at an upriver angle against the current. But they were headed for the opposite shore and survival.

Behind us the roars and bawls and gasps of the cattle, the loud clang of the coal-oil tins, and the wild yells of the boys rose high and distinct above the gushing roar of the river. I looked back over my shoulder and saw the mass of cattle surging through the funnel in a great, unbroken, brown-and-white line.

And then it was all over. The last cow came gasping up out of the swim and headed for the scattered piles of hay.

We made a quick tally. Every single critter that had trailed out of the Batnuni Ranch so many agonizing days before was here, safely across the Blackwater River, and would soon be working on the mountainous piles of hay at the Pan Meadow.

I yelled at the boys, "Nice going, you guys! There's not many cow outfits would have landed this drive!"

Simrose grinned at me from the back of the steaming, ice-caked Big Enough.

"Hell, man," he said, "that wasn't too tough."

Baxter laughed.

Jimmy John waved a mittened hand at us.

"More better you boys don't get too proud. I think about it long time: Great Spirit—He must have been riding herd on this frozen drive."

CHAPTER X

———❦———

Development of a Swamp Ranch

THE PAN MEADOW was a large-scale swamp. Slough grass, wild redtop, bluejoint and Reed's canary grass sprouted here and there across its wide expanse. The Pan Meadow country was much like Joe Spehar's original swamp and parts of the Chilcotin and Anahim Lake ranges, where mud bogs, shallow lakes, floating islands, and air-exposed muskegs form a good percentage of the land surface.

Thirty acres of dry, open grazing land in southern British Columbia and the northern States can graze one cow. In the semi-desert and dry regions of the Southwest, anywhere from eighty to two hundred acres are needed per animal. But an otherwise useless swamp when properly drained, burned and flooded can range a lot of cattle. A cow can fatten on from one to three acres of this land, and one of the happiest results of a drained swamp is that the more the land is grazed and worked over, the better the grass becomes and the more palatable the new green sprouts are to the stock.

Since my book *Grass Beyond the Mountains* came out in 1951, I have received many letters and requests from all parts of the States and Canada for information on the swamps and ranges of British Columbia. Unfortunately I don't seem to be able to squeeze in the time to give each of these interested folk the answers.

For those who are interested in this new type of cattle ranching, I will try to give a brief round-up of what we did to put these Pan Meadow swamps—a typical swamp ranch development—into production.

At the Pan Meadow—when we first discovered it— two small, mud-bottomed creeks merged into a long, dangerous slough where several springs bubbled up out of the mud. Along the shores close to the water's edge, where there was good drainage, wild hay was luxuriant. The further back you got from the slough the less the grass grew, and the more boggy the land became—and yet the peat and soil patches were the same type as on the better drained ground, showing that drainage was all that was needed.

Wearing hip-length rubber boots and using ordinary foot-pushing hay knives, two men cut out blocks of rooty sod two feet square, in a long straight line. Another man, using a pitchfork with bent tines, hauled the sods up on the bank. Our first drain ditch on the Pan Meadow was a quarter of a mile long, two feet wide and two feet deep. The three-man ditch crew made this quarter of a mile in three days. A small brook soon flowed down the ditch where there had been no water flow before.

The first year of the development at the Pan Meadow I contracted the ditching with Peter Morris. I paid him a hundred and fifty dollars a mile. Peter's crew dug two and a half miles, and Peter claimed he made money on it.

Jimmy John, as Indian foreman, insisted that the

drain ditches be done first, before any other development. Peter put six Indians to work on the ditching and
paid them one dollar a day per man. Small log-and-rock
head gates were built at strategic points along the creeks
for spring flooding.

After the swamps were drained, an early spring fire
was essential. The fire was set while the frost was still in
the ground. The grass roots were not killed and forest-
fire hazards were avoided. After the burn, the black ash
on the ground added important elements to the coming
new growth. If the same fire was set either in the summer
or fall, the grass roots would be killed and it sometimes
takes years before a new sod forms.

Three weeks or more after the burn, the meadow
was flooded by dropping in a head gate at a crucial spot;
or sods can be thrown into the lower end of the ditches.
The surface water or sub-surface water was left in the
swamp from about May twentieth to June fifteenth and
then released.

This burning and flooding system causes a vast growth
of the strongest in protein value, the most palatable to
stock, of all wet-meadow grasses, wild bluejoint, to spring
to life—averaging two tons of hay to the acre. If the water
is not turned on the meadow in the early summer, slough
grass and a species of volunteer redtop will take its place,
and if the main bulk of the hay is redtop, you can usually
count on only one ton of hay per acre.

Some meadows are hummocky and too rough to cut.
If a heavy duty disc or a powerful iron-tooth harrow is
drawn over the surface, the hummocks are scratched and
torn up, leveled down a bit, and the air and sun gets a

chance to hit the newly exposed earth. It is miraculous how all types of wild grasses shoot up suddenly out of hiding.

Straight slough grass or rip gut, as some farmers call it, varies in feed value from meadow to meadow. It has very little strength in it when cut as hay; but it is the rancher's gold mine when it has been burned off in the spring and is used late in the fall and winter as pasture. Stock will eat slough grass almost any time of the year, and despite the fact that it is considered soft feed, they will stay in remarkably good shape on it.

A thirty-by-eighteen-foot, two-room log bunkhouse was contracted for at the Pan Meadow for two hundred dollars, with broad-axed floor and shake roof. The Frontier Company furnished the doors, windows, nails and hinges.

A ten-horse, twenty-eight-foot-square, log barn with dirt floor and high-shaked roof with haymow was contracted for a hundred and fifty dollars. Smaller log buildings such as a meathouse, a supply cache on stilts and an outhouse were constructed at thirty dollars per building.

We freighted in three heavy-duty mowing machines, four hayrakes, harness for seven teams, and chains, double-trees, quarter-inch cable, half-inch sling ropes, prospector's forge, eighty-pound anvil, and the other equipment necessary to put up hay.

To turn the Pan Meadow swamps into a series of lush meadows and green pastures, plus buildings and equipment, cost us forty-four hundred dollars, but it must be remembered that all of this work and the purchase of machinery took place in the late thirties.

It took three years before the Pan Meadow and its adjoining swamps were in full production. The fall and winter rustling was unlimited, the grass had improved so in palatability and nutritive value that stock could usually rustle out in good shape until after Christmas, be fed for three months, and turned out on the range early in April.

Drainage and early spring burning had greatly discouraged the flies and mosquitoes, the shaking bogs and dangerous mud flats had settled, were producing grass and were safe for stock. The Pan Meadow development proved that these marginal and useless lands can be put in shape to more than quadruple the original production.

I want to point out here to the countless number of hopeful ranchers and meadow hunters who are interested in this country that the Pan Meadow, with its more than fifteen hundred acres of drained swamps and muskegs and natural meadows is located some hundred and fifty wagon or sleigh road miles from the nearest railroad or town.

Although the interior of central and northern British Columbia is dotted with unclaimed, unsurveyed, wild-meadow possibilities similar to the Pan Meadow, it is necessary to go back into the bush with packhorses, and lots of time, to locate them. Nowadays the meadow hunter is helped immeasurably by Air Force and government air photos. British Columbia has been systematically photographed from the air in the past few years and maps and air photos can be obtained from the Parliament Buildings in Victoria for a reasonable amount.

Naturally most large-scale swamps and meadows located within striking distance of railroad and highway have been snatched up by settlers in the last few years.

Therefore, unless a person is prepared to buy an existing ranch, he must realize that it takes time, money, know-how and a fair amount of stamina to plunge into the bush in search of a meadow and stay with it until he has developed it.

When an outfit plans to settle on British Columbia's vast, little-exploited frontier, it quickly finds out that the government departments are unique in this day and age of red tape and long, time-wasting negotiations. Considering the Province's gigantic territory, a land area so immense and so diversified in its climate and natural resources, where the combined areas of Washington, Oregon and Idaho could fall within its boundaries, it would be presumed that any business connected with government lands, forests or agriculture would be difficult to close.

But this is not so. Acting for the Frontier Cattle Company, and, later, for Pan and myself in the matter of lands and livestock, I have had to make a number of trips to the Parliament Buildings in Victoria, the capital of British Columbia, and each time I have cleaned up my files within two or three days, and reluctantly left this beautiful city, with a warm glow of appreciation for the personal attention and the quick and easy settling of my affairs by the various government officials.

I attribute the great efficiency of the Departments of Lands and Forests and the Livestock Department to the excellent British Columbia Lands Act and the outstanding men who administer it; such men as Frank Morris, the Surveyor-General of British Columbia, who came up from a rodman, then a clerk, to the head of his department; and the men who work with him: the former Deputy Minister of Lands, George Melrose; the Superintendent of Lands, Robert Burns; the great pioneer surveyor, Swannell; and the fabulous packer, Skook Davidson, and the dozens of other men who have organized and put down on paper the vast potential of this, until recently, little-known forest empire.

The Department of Agriculture and the ranchers and farmers throughout B.C. are extremely fortunate in having as the head of the Livestock Branch, Dr. Wallace R. Gunn, whose wide acquaintance with the bloodlines of different breeds of cattle and horses almost equals his knowledge of livestock diseases and range management. Dr. Gunn spends many hours of his crowded schedule setting up new ranchers, and his efforts on behalf of the Frontier Company and other cow outfits of our acquaintance were largely responsible for their sound practices and consistently improving ranges and hay lands and cattle herds.

CHAPTER XI

Killers of the Backlands

THE FIRST PROBLEM that confronted Simrose and me at the Pan Meadow after the boys had all ridden off—Baxter, Aitkens and Stobie to the Home Ranch units to help Pan, Jimmy John to his home in the valley between the Fawnies and the Nechako Mountains, Ed to the Striegler farm at Vanderhoof and Rob to the Batnuni—was the task of cutting out the one hundred head of cattle for Lashaway and moving them to his hay ranch. The road from the Blackwater crossing to Kluskus Indian village took a wide-fronted horseshoe route to reach Lashaway's. This trail drifted across some eighteen to twenty bush miles. By plotting the approximate position of the Pan Meadow on the reconnaissance map of the area, Lashaway's appeared to be not more than five miles away. A series of jackpine hills that rose to small, heavily timbered mountains reached off to the south in the right direction.

Simrose and I decided to ride through the hills and look out a possible cattle trail. One morning we opened the water holes and fed the cattle early, then struck south through the bush. Our saddle horses picked their way cautiously around some muskegs, bypassed several steep banks and wooded hills, then climbed up onto a high, sparsely timbered hogback that ran in the right direction.

It was easy traveling along the ridge. Before we had realized the distance we had covered, we rode out onto an old, stump-sprinkled, wagon road that bordered a wide, slough-grass swamp. We rode abreast down the road a short distance and, rounding a bend, saw before us a good-sized, two-story, weather-grayed frame building, several smaller log cabins and outbuildings, a barn, corrals, and cattle shed.

Several sleek, short-coupled, black Percheron horses were feeding behind the gate fence. They threw up their heads and watched our approach. One of them whinnied—the front door of the frame house flew open, and Sundayman Lashaway stepped out, looked at us for a moment, then smiled and waved his hand.

"Hello, hello," he said.

Simrose and I dismounted and shook hands with the Kluskus Indian. That is one white-man custom the Kluskus and Nazko Indians adhere to—a handshake with gloves off is an understood formality when friends meet.

Sandyman or Sundayman Lashaway was distinguished-looking and well put together, perhaps forty-five years old. He stood about five-ten, was heavily built, and, unlike most Kluskus Indians, he had a comfortable, prosperous look about him. He wore an expensive 3X beaver Stetson, frontier pants, custom-built riding boots and homemade, beaded moosehide jacket.

Lashaway maintains the stoical aloofness of the Indian until he knows you well enough to call you friend. Then his broad smile and happy laugh are infectious. The name Lashaway was used with awe and respect by Indians from Quesnel to Ulgatcho. He was what the Indians call a Rich Man and a smart man. But to my notion Lashaway was more than that, he was a dynamic man who packed tremendous energy.

At that time he ran nearly a hundred head of top white-faced cattle, and forty or fifty head of well-graded-up Percherons. Lashaway farmed his huge trap line rather than gutted it. When fur prices were good, he and his family made thousands of dollars in a season.

A group of down-and-out Indians and their families always hung around Lashaway's. He put them all to work. Lashaway gave each individual the kind of job that suited him most. In return for their work he furnished them with flour, sugar, tea, salt and ammunition.

I told Lashaway about the short-cut route to the Pan Meadow. He did a bit of mental figuring and made me the astounding offer of only fifteen dollars a mile to slash a cattle trail twelve feet wide, connecting his place with the Pan Meadow.

"And then I help you drive those cows over," he said.

We fed our horses and, with Lashaway in the lead, walked towards the house. There was only one person and little furniture in the big, whitewashed, living room. Lashaway held open the door for us to enter, then stepped in behind us. A neatly dressed, middle-aged woman with red hair (who I later found out was half white, half Indian) looked questioningly at us from a chair behind a foot-pedal Singer sewing machine in the far corner of the room.

"My friends," said Lashaway. "This is Rich Hobson—and this Simrose."

Immediately the woman's face changed. A pleasant smile, hospitable and warming, replaced her puzzled look of a moment before.

Simrose and I stood around, stupidly juggling our hats, while Lashaway crossed the floor to an old-fashioned china cabinet. He picked out three intricately designed glasses and set them down on a white, newly painted dining-room table.

"Maybe you boys are cold," he said. "I think hot rum make you feel better."

I said, "There's nothing the matter with Simrose here, but I think I'm cold sick."

The following day Lashaway, his nephew Little Thomas, and several other Indians started cutting out the cattle trail. They put the trail through to the Pan Meadow in four days, then helped us drive over the hundred head of cattle that Lashaway had agreed to winter for us.

It was about three o'clock one morning, back at the Pan Meadow, when Simrose and I were awakened by the frightened bawling of the cattle and the strange, wild, gutteral growlings and snarls of a pack of wolves. It was a bright moonlit night. We threw pants and coats on, grabbed rifles and ran down the incline to the milling, bawling herd.

Big, shaggy-haired wolves were dashing here and there among the yearlings. Despite the moonlight on the bright snow it was hard to distinguish between the yearlings and the wolves.

"Shots in the air," yelled Simrose.

We scattered the wolves and the cattle in every direction with a volley.

The wolves disappeared in the brush around us. We stood night guard for at least two hours among the cattle who finally quieted down. It was one of the few times I have seen a disorganized wolf attack. Usually these killers circle a bunch of cattle until the critters are wild with fright, milling around in a close circle. Then one of the wolves dashes into the outside edge of the bunch and cuts one cow away from the herd. The rest is easy. The cow is hamstrung and devoured. On this particular occasion, three yearlings lost their tails. That was all.

A day or so later, Lashaway rode over to report that a big pack of black-and-brown wolves had milled the cattle on his meadow and killed one of our yearlings.

We rode over to Lashaway's and spent the day on top of a high haystack in the middle of a big, open meadow waiting for the wolves to reappear at the kill. Just before dark the pack came boldly out of the timber and trotted across the meadow towards the half-eaten yearling. We got several shots in from our hideout but we didn't hit one animal. The evening light was not too good and the targets too far away. We counted thirty-two wolves in this pack.

This was the beginning of our wolf troubles. For years we carried on a running battle with the northern wolf.

These fierce-looking, shaggy-haired killers have bearlike heads and long, hooked fangs. As far as I can find out they are the largest and most ferocious canine breed in existence. They have seeped down out of the Yukon and Northwest Territories into central British

Columbia and Alberta during the past twenty-two years.

Men who live in the backlands estimate that during the fifteen years previous to 1950, when the British Columbia government went to work on the wolf packs with planes and poison, northern wolves killed, crippled and worried to death a frightening percentage of the moose population in central and northern B.C. The once-great caribou herds were hit very hard, too, and deer and mountain sheep were eliminated altogether in some areas. For a while the wolf threatened to exterminate our big game, and nearly wiped the frontier cattle herds off the range.

The northern wolf, quite possibly a cross between the Siberian and the black Alaska wolf, runs in three colors—black, brown and gray, with the black becoming more predominant each year. Long, wiry hair adds to the impression of his enormous size.

The northern wolf's head is large, wide between the eyes, tapering to viselike jaws with the long, hooked fangs. These wide-backed jaws can lock together, and the hooked teeth hold hard and fast to rip a horse, a cow or a moose from stem to stern. The jaw can unlock into a barrel-like neck, thus allowing him to gorge down more than ten dollars' worth of T-bone steak in one short swallow. The animal's bull neck looks almost as if it were a part of his deep, massive chest.

But the size of the wolf's heart is even more astonishing. Anyone who has examined his heart, his great neck and chest, sloping gracefully to a whippetlike sinew-and-bone hindquarters, and who has seen the killer in action, can easily understand the terrific feats of strength and endurance attributed to him.

Once I saw where a single wolf had dragged a six-hundred-pound, yearling steer at least twenty-five feet, across rough ground, to get the carcass clear of some of our steel traps.

Several years ago, when only a thin crust covered a three-foot depth of snow—a condition which according to bunkhouse talk gave the moose a decided advantage in high-trotting away from the wolf packs—I came upon a series of deep holes in the snow. They were about eight feet long and three feet wide and all ran in one direction. In between them, I found a deep swath cut through the snow by a running moose. About two hundred yards further on I rode up on what remained of a big bull. I was amazed at the evidence revealed in the snow. The holes had been made by wolves. They must have been moving at great speed, springing twenty feet at each jump and not having to reset themselves to make the next leap.

The northern wolf dwarfs the American and southern wolf in size and bone structure. The few timber wolves, or lobos, that still roam the Rocky Mountain States and the southern deserts don't average much more than eighty pounds in weight, with the largest animals reaching up to a hundred and twenty pounds. Their necks and jaws are smaller, and their heads resemble the coyote rather than the fierce, almost bearlike, head of the great Canadian northern wolf.

I can find no records of northern wolves in central B.C. until 1928. That year a pack of seven big black wolves was first seen by Captain Alexis, of the Kluskus Indians, near the spot where we threw together the Home Ranch. Since that time more packs have drifted down from the north and apparently have bred rapidly.

Timeworn legends and more recent reports from northern Siberia, before Iron Curtain days, speak of an annual toll of humans killed and devoured by half-starved wolf packs; but I can find no official record of anyone being killed by wolves in this country. No doubt a few prospectors and woodsmen have been eaten by wolves, but there was no evidence produced to indicate that they were killed by these animals.

Unless they are very hungry, wolves have an instinctive fear and distrust of man smell. Also, the great northern wolf, to date, has had enough feed from game and rodents to keep him fat. But they have given more than one person a bad scare.

It is hard to convince Clayton Mac of Anahim Lake, for instance, that the dozen wolves that surrounded him on a knoll were circling him out of friendship or curiosity.

Clayton had killed a moose near Anahim Lake and was skinning it when he heard a wolf howl. Other howls rose up out of the jackpines surrounding the small opening. Clayton's horse let out a wild snort, broke his hackamore shank and stampeded for home. Clayton straightened up from his work in time to see several wolves standing at the edge of the clearing. He had three shells in his gun. He pumped one of these into a big, gray wolf and it dropped over dead, but the other wolves scattered into the brush for only a moment, then broke back into the opening, running and howling.

Clayton thought it was the half-butchered moose that they wanted. He slowly withdrew from the carcass to a small, open knoll. There were eleven wolves, mostly blacks, in the pack; they were led by a big female. She took them right by the moose and began closing up the circle around the knoll. Clayton looked quickly towards the timber and estimated that it was at least a hundred yards to the nearest climbable tree.

He did not notice the big black bitch wolf until she had slithered up to within six feet of his rear. She was flat on her belly, great unblinking eyes staring up at him. Clayton let out a yell and swung his gun butt with terrific force at her head. He said she avoided the blow like a boxer.

As he started another swing with his rifle he glanced over his shoulder and with a quick pivot let fly at another wolf which was lying flat on its stomach behind him.

"I couldn't possibly hit those wolves," Clayton told me. "They just watched the gun butt and moved back a little to let it go by, then came in a little closer.

"The pack was closed in behind the two lead wolves and I began to say the Lord's Prayer when my partner, Sam Moody, rode out onto the flat, shooting his six-shooter and hollering like a wild man. When the wolves saw Sam and the buckskin come exploding out of the bush like a war turned loose, they stood for a moment, then slunk away—but Sam got one of them.

"If Sam hadn't popped up when he did I guess they would have claimed I died of a heart attack before the wolves had made a meal of me."

Clayton Mac and Sam Moody showed me the skins of the two wolves.

This sort of wolf episode doesn't happen very often, but then the big northern wolf is most unpredictable.

You just can't guarantee what his behavior is going to be at any given time or place.

An Indian named Jiggy Jim took over the Frontier Cattle hay contract one summer on the Pan Meadow. One day while Jiggy and his crew were out haying, a pack of wolves came howling out of the timber in the direction of camp. Mrs. Jiggy saw them coming, grabbed up the baby and herded the other small children into the family tent. The wolves were in hot pursuit of the four camp dogs.

Mrs. Jiggy saw that their black collie dog wasn't going to make it. Grabbing up the only weapon available, a single-shot 22, she let fly into the pack. But it didn't help the collie. His helpless cry was smothered beneath the hairy mass of on-rushing wolves.

The three remaining dogs crashed into the tent. The children were screaming and crying and hanging on to their mother's dress when she spied a big pile of blankets. Under the blankets went Mrs. Jiggy and the children and into the tent charged the wolves.

It is easy to imagine her fright as she lay with her children under the blankets with the savage bunch of howling, fighting devils tearing the dogs to pieces on top of her.

Later Mrs. Jiggy said to me, "I pray to the Good Father to save my little girl, and then I hear lots of voices and shooting and Jiggy hollering like crazy man—and then tent he fall down."

I inspected the bloodstained tent, the hide of the one black wolf Jiggy got, and saw the chewed bones of the two dogs killed and the wounds of the others.

These northern wolves mate only with one individual, and they are loyal to their family. They are never bitten up during mating season. The pups are born around the first of May in wolf dens—deep holes or small caves—and that's when the females are stocking meat ahead. In the late summer and fall, when the bitch and dog are teaching the pups to kill, is the other time when their list of victims rises. They have an average of six pups to the litter.

The wolf family is a close-knit corporation. The whole family stays close together except when hunting. After the family is grown they'll spread out fan-shape through the bush, driving game ahead of them.

Wolves will kill differently at different times of the year. In the summer the packs kill on the run, and it's fast business. But in the winter they wear the game down by degrees. When the victims are played out the wolves leap across the snow and hamstring them.

They start eating, working from the hindquarters forward. They'll eat from twenty-five to thirty pounds each to get a real fill-up, then slink back in the bush and lie down for a rest. Sometimes the killers allow the game to live just as long as it can, often going back to eat on the animal for two or three days before it dies.

Despite the northern wolf's brutal killing methods and even though he nearly put our outfit out of business on several occasions, I still have a great liking and respect for him. Strange wolf howls floating up out of the forest still give me an inexplicable feeling of exaltation—a feeling that the great, untouched wilderness of British Columbia is still unconquered by man.

Sometime after the Frozen Drive (I abandon strict chronology in order to continue the subject of wolves) a Stony Creek Indian named Mike Ketlo gave me a present of a male wolf pup. The old Bear didn't like or respect the scent of this pup, but in his high-toned and seemingly uninterested sort of way he got along with the Wolf until the animal was nearly a year old and twice the size of the Bear. The jealous Bear resented the Wolf more and more; chased him out of the house and one day made an almost fatal mistake.

He grabbed the half-grown pup by the scruff of his neck. The Wolf had never before tried to defend himself. Now I saw the difference between a dog and a wolf when it comes to fighting.

The Wolf rolled over on his back, broke the Bear's grip, opened his long, wicked-looking jaws and the next thing we knew the Bear was writhing in agony—his head locked inside of the Wolf's mouth. We threw pepper, buckets of water and broom handles at the Wolf, trying frantically to unlock his jaws, but to no avail.

Finally, just as the Bear was going limp, I lay down on the floor next to the struggling animals, got a scissors grip on the Wolf's body with my legs and pried the animal's mouth open with an iron wrecking bar.

There were two holes in the top of the Bear's head, one over half an inch deep, and three jagged holes in his under-jaw. He was sick for days afterward.

Once on a cattle drive the Wolf heard me trail-talking to some balky steers. He appeared suddenly out of the bush, and in one terrific leap landed on a steer's back. The drive stampeded in every direction. Of course the

Wolf was only trying to help us, but if I'd been packing my six-shooter at the time I'm afraid I'd have plugged the old boy.

The Wolf was fascinated by the sound of horse or cow bells. He spent a lot of time following, and seemed to have a very high regard for, any belled animal. Simrose scared up a small, high-clanging sheep bell and collar and put it on the Wolf. Wolf was very proud of this bell and he shook and rattled it, with his head cocked to one side.

Simrose composed a song which the Wolf loved to hear. He practically danced to the monotonous tune which we chanted at him:

"Here comes the Wolf Wolf Wolf,
Here comes the Bell Bell Bell
The Wolf Wolf Wolf and the Bell Bell Bell."

This bell was a good thing, too. We didn't put too much trust in the Wolf and his clandestine activities. Some day we feared he would go to work on a beef. But the bell warned the critters that something strange was coming up on them. At the same time the Wolf got a feeling of comradeship with cattle. The Wolf and a belled cow had something in common.

When the Wolf was a year and a half old he fell in love with another wolf song. A high, fluty call rose up out of the Pan Meadow forests. The Wolf pricked up his ears, but resisted the temptation. The female wolf hung around the hay camp for several days—making the saddest and loneliest noises I've ever heard—and, of course, the inevitable happened. The Wolf couldn't keep up his resolve to stay clear of women any longer. I must say I couldn't blame him.

The bitch wolf's final and most enticing call broke him down, and the Wolf dashed into the jungle with his bell ringing wildly. He was gone for nearly a week, and Jimmy John swore up and down that the other members of the wolf pack had caught up to the female and our Wolf, and killed him. But late one night he arrived back, a messy, beaten-up and chastened fellow.

The final wind-up with my wolf pal came when he jumped another steer, enticed a wolf pack into the neighborhood again, and just about killed the Bear in a second engagement. I knew then that I would have to get rid of him. Before long he would be killing cattle; and it was just a question of time before he did away with the old Bear.

George Alec, a kind, young Indian lad from Trout Lake, urged me to let him take the Wolf. He could use him on his trap line as a pack dog. Unlike some Indians, George was good to animals. I knew it was the only out, so, with tears streaming foolishly down my face, I watched George Alec lead him by the collar down the trail.

The Wolf kept turning around in the trail to look at me. His eyes had a questioning and even tragic look. I turned my back to him and walked in the opposite direction.

For the following two years I was able to keep track of him. He became a great cougar dog—a top trap-line companion to George. But he was a dog killer and a watchdog that bore watching. The older he got the fiercer he got. The Indians had to keep him chained up at Trout Lake. No one but George could go near him.

One day I pulled into Trout Lake with a freight team. This was one of the few trips when I had been able to get clear of the ranch without the Bear.

At Trout Lake, George Alec's brother, Felix, had offered his new, clean, one-room cabin to me for the night. The Wolf was chained up near the door. I yelled hello at him. He had grown since I'd seen him last. He stood up on hind legs against the chain, his yellow eyes had a strange, wild look to them. I crawled down off my wagon and walked towards him. Deep, rumbling growls rose out of his throat.

The Indians all began yelling at me: "Stay away. He'll kill you if you get close—Look out, Rich. Don't go near him."

I walked slowly and deliberately up to him. When he was on his hind legs his great set of gleaming teeth reached slightly above my throat. I shucked loose my first feelings of fear. I threw my arms about him. His great paws encircled my neck and his long, dripping tongue licked my face. For a long moment we hugged each other, and I could feel his affection reaching out to me through his warm body. I got permission from Felix to turn him loose that night and take him into the cabin with me. His weird, yellow eyes followed my every move. I talked to him as of old, and he talked back in his throaty sort of way. The Wolf snuggled against my back most of the night. An hour or so before daylight I crawled out of my blankets and lit the coal-oil light.

I boiled up some coffee, threw the Wolf a hunk of bread and began explaining things to him as best I could—telling him why I had to leave him, and that it was the only thing I could do. Finally I crawled back into my bed.

When I woke again at daylight, the Wolf had disappeared into the forest. He had jumped through the open window.

I never saw him again.

George Alec saw him twice afterward. Lashaway came upon him several times. Each time he was sighted it was in the vicinity of the Pan Meadow, nearly sixty miles beyond Trout Lake. That is where he would have memories of his puppyhood—happy, carefree days; and the Pan Meadow was where he had met his first love.

Several years later the Indians of that country put on a great drive to eliminate a fierce pack of killer wolves in the Pan Meadow area. The Wolf died in a wire snare four miles this side of the Pan Meadow crossing. Lashaway sent the word over to me. He knew I loved that Wolf.

Now we have only one pack of wolves in the big sweep of forests that surround Rimrock and River Ranches. They are not cattle killers. I want them to be protected in this private game sanctuary of our ranches, just like all the other animals.

CHAPTER XII

The Red Line of Death

I WAS SPENDING SOME TIME at Batnuni Ranch, helping Rob Striegler work over our hospital bunch and haul in extra hay and firewood, when a strange urge to get back to the Pan Meadow began to work on me. I wasn't supposed to land there on my long circular ride—Batnuni to Pan Meadow to Lashaway's to Nazko—for another ten days, but I suddenly changed my mind and decided to ride immediately to the Pan Meadow. The strange, back-lands moccasin telegraph, for which I hazard no explanation, had connected.

I got away to an early start. The weather was mild. My saddle horse Black Bear leaned into his hackamore bit. The dog Bear fell into his usual lead position and we drifted across the uncut miles of windfall on our short-cut, saddle-horse trail in record time. We had good luck at the Blackwater crossing. Ice had at last closed over the roaring torrent, as it nearly always did late in January or early in February.

I tied my horse to a tree and, using my saddle axe, tested the strength of the ice. Two and a half inches of ice which rests solidly on flowing water with no air space between the two, can safely support a man and saddle horse. Four inches will support a team and sleighload of two tons.

There are all types of ice. Clear ice stretching across deep, black-looking water on a level lake will support three times the weight that medium river-ice can stand.

We always test the ice up here before crossing any body of water, for many teams and a number of men have gone to a watery death because of over-ice-confidence, as it is referred to in the bush. In order to distribute the weight over a larger surface and be in a better position to move fast if the ice gives way, a person should always lead his horse.

The Bear dog made his own test of ice by crossing it. I watched him. He held his shaggy tail high in the air, swinging it from one side to the other in a kind of balancing motion. Several times he skidded to a stop, lowered his broomlike tail, cocked his head to the side and listened intently to the roaring noise made by the river as it hurtled by beneath the three questionable inches of medium ice. The old Bear would straighten up his tail after he was satisfied and step cautiously ahead for a few steps, then break into his shuffling trot.

At the far side of the river, the Bear sniffed at a big spruce tree for a moment, then turned round and looked back at me. When he saw that I was waiting for his final decision, he galloped nonchalantly back across the channel, this time paying no attention to the ice. He came up smiling and with his tail wagging.

I scratched his ear, gave him a pat and, untying my horse, we advanced across the river. The ice made a slight swaying bend when we got over the middle of the channel. I was glad when we reached the spruce trees on the far side.

It was nearly dark when we broke out onto the wide Pan Meadow opening. Far up ahead I heard cattle bawling, which meant that something was wrong. It was black dark by the time I reached the corrals and barn. Cattle were milling about an unloaded rack of hay. They were hungry. The load of hay had been almost cleaned up by the bunch that surrounded it.

The team had been unharnessed and turned loose with the saddle horses and a gate left open so they could get out onto the meadow to rustle.

I looked up towards the cabin. No light.

Everything pointed to a last-minute adjustment that a good cattleman would make if he wasn't going to be around for several days.

I hurriedly stabled and fed my horse, shoved through the cattle, and trotted up the hill to the bunkhouse.

I pushed open the door and yelled, "Simrose," into the darkened room. For a moment there was no answer and then I heard his voice. It was weak and crackly.

"Rich—that you? Lamp's in corner by stove—"

The last of his words was strained—barely audible. I didn't ask any questions. I groped around for the lamp. I heard Simrose retch violently. I found the lamp and lit it with my lighter—then crossed the floor to the bedroom carrying the light.

The dull glow of the coal-oil lamp flickered weakly over Simrose's gray face. He was vomiting over the edge of his bunk into a lard pail.

I wiped his face off with a towel.

"Blood poison," he gritted. "Scratched my hand on a rusty nail a week ago. Tried to make Lashaway's this morning. Fell off my horse—couldn't make it. Been down four days now. Don't know if you can do anything—"

One of Albin's hands was wrapped in a wet towel. He pointed at it.

Now I set the lamp down on a coal-oil box next to the pole bunk. I stood over Simrose and carefully unwrapped the wet towel.

His hand didn't resemble a hand. It was swollen and shining like a balloon. The back of it, near the wrist, was black. I turned it over and looked at his arm, hoping against hope that I wouldn't see it—but I did—a dull red line running up the inside of his arm to his armpit where a hard little knot had formed.

I tried to sound nonchalant.

"O.K., boy," I said. "We'll fix you right up. Three days and you'll be up and about. I'll be back in a minute—got to start the fire and heat up some water."

"Sorry about the cattle," Simrose whispered. "Got hay out to 'em till yesterday. Got water holes opened this morning, but I couldn't pitch hay—head aching like hell—kept passing out. Should have tried Lashaway's yesterday."

"You've done a great job as it is," I told him. "The cattle have only missed a feed or two. Don't worry about that."

I left the room, headed for the woodpile, got a fire going in the cookstove and filled the empty water buckets with fresh snow.

While I was lining things out I thought back over the years to blood-poison experiences that had been thrown in my direction. I knew I couldn't afford to make a mistake. There wasn't much time left—every wasted minute lessened Simrose's already slim chance of pulling through.

He was in the second stage of blood poison, and was now approaching the third and last stage. The line of blood-poison death was easily determined by the red line that ran up the inside of the arm to the armpit, where the glands swelled, or inside of a man's leg to the groin where a similar congestion would take place. The frontiersman's term, the red line of death, couldn't have been a more fitting description, for it meant that the poison had entered the blood stream, and in those days, before the wonder drugs of sulpha and penicillin, unless a deep-cutting operation could be performed, the victim had little, if any, chance of living.

The knot under Simrose's arm meant that in a short time the poison would reach the heart. Then it would be all over. Quite often between the second and third stage of blood poison, amputation of the affected part is necessary to save the life of the victim. That is where a layman's skill ends and a surgeon is necessary. He has to decide just how far up the red line to amputate. Sometimes a second and then a third amputation is necessary, as the surgeon moves ahead of the storm center of the blood-poison pressure, surging forward towards its ultimate objective—the heart.

The Pan Meadow was over a hundred and fifty sleigh-road, deep-snow miles from Quesnel and the nearest doctor. It would take five days with a grain-fed team to reach it. Simrose would be in the far-beyond long before we reached medical aid. It would take four changes of

relay horses and at least forty-eight hours to reach
Quesnel by saddle horse to get a plane, and, what's more,
it was doubtful if any plane, in use at that time up here,
could have safely landed and taken off at the hummocky
Pan Meadow.

There was only one chance, and a mighty slim one at
that, to pull Simrose through. I would have to operate
on him myself, and put my trust in God that I would do
the right thing, for I was fully aware of my inadequacy.
If Simrose lived through this ordeal something greater
than just Rich Hobson would be responsible.

Now, as the snow melted down in the pails and the
first bubbling sounds of boiling water spread through
the quiet room, I began my preparations.

There was a high, flat, broad-axed bunk in the kitchen
end of the cabin, and next to it sprawled a low plank table
used as a kitchen workbench. This would be the best
operating room set-up.

I cleaned off the table and pulled it up to the bunk,
then unrolled a wide chunk of clean gauze and tacked
two lengths of it across the table where I planned to strap
Simrose's arm down. I scalded out an enamel wash basin
and an aluminum cooking pot, and, as I worked, I tried
to visualize what was soon to take place.

Our first-aid kit consisted of little more than mercurochrome, adhesive tape, bandages, a small operating knife
that I had used for cutting calves and lancing cattle infec-
tions. I had several pairs of pliers and a half-dozen, unused
razor blades. One pair of long, narrow-nosed pliers was
made to order to hold a razor blade with. I taped the nose
of the pliers tight with the razor blade between its teeth.

This would give me a good handle to grip when I was doing the cutting.

I rolled several slivers of gauze into thin little rolls to be used for drains.

Now I noticed that the dull light of the coal-oil lantern was so dim that my eyes hurt when I tried to concentrate on a tiny object for long. It occurred to me that candlelight would throw a softer, more natural light around the table. We had a stock of ordinary candles in all of the company bunkhouses. I found a box of them, plastered six down solidly in convenient spots around the table, and lit them. My light problem was now solved. The candles threw out a soft but penetrating glow—far better than the coal-oil lantern.

I refilled one bucket of snow water and scalded my crude instruments in the aluminum pot. I set a pair of pliers in a convenient position close to the pot where I could use it to lift them out.

Now I arranged small piles of cotton, different lengths of gauze, and the quart bottle of mercurochrome on one end of the table where they could be reached without undue fumbling or misplacing. I brewed up some powerful tea, sipped down a half-cup myself and took a full cup into Simrose.

I returned to the operating table and once again inspected every little detail of that workbench. As I looked over this crude backwoods setup and drank another cup of tea, I tried to make the final decision. In the next few minutes I would have to make up my mind and then put everything I had into the course that I chose.

I remembered that in the old days, rather than take a

chance with a man's life, many doctors performed an amputation. But amputation by an experienced surgeon with the proper equipment and facilities at hand was a far different proposition than it was here in the remote Pan Meadow with me as the doctor.

The thought of amputation appalled me. My head swam. My mind drifted back through the years to the East Texas oil fields—the Discovery Well, Longview, Kilgore, Gladewater, at the time the toughest towns in America. In the great oil-field rush, overnight, hundreds of oil derricks were flashing high into the sky across the great East Texas newfound oil dome. The date—1931.

I was working as a roughneck for an oil-contracting outfit who used old and obsolete equipment, but banged through a producing well every twenty-six days, at thirty-eight hundred feet to Austin chalk and then oil. Old-time oil drillers and roughnecks from East Texas will remember the flimsy, wooden derricks, the giant, back-breaking, hand-swinging dun tongs—rotary rigs—worn-out safety belts—oil-well tool supplies—the last gasp of dangerous, outdated equipment, even at that time shunned by the big oil companies, who were using cable tools, shorter hours, safety devices and the many labor-union precautions.

A picture flashed across my mind of the big, barnlike, newly thrown-up, lumber hospitals—loaded to capacity, overflowing with unprecedented numbers of oil-field casualties—inexperienced interns replacing much-needed doctors as they worked day and night to save the lives of the unfortunate roughnecks who had ventured to work for higher wages on obsolete equipment.

I thought of the day when, swinging my set of dun tongs through a spray of flying mud into the whirling pipe above the rotary rig, my glove got slightly fouled, was torn off my left hand, leaving a slight, not-worth-mentioning cut. And then the surprised look, followed by agony, written across the face of the back-up man, the roughneck working opposite me, when he looked at his hand—crushed when the traveling block broke loose.

I remembered the infection that took hold of my hand several days after the accident, the lancing of the knuckle followed by a further and unexplainable swelling.

I thought of the free-for-all fist fight in the roaring tent city east of town when I joined my fellow rough-necks in an all-out, clean-fun, supremacy battle with a neighboring crew. I could not easily forget the terrific pain that shot through my infected hand to my shoulder blade when I connected with somebody's head. And then—a deep throbbing—a red line running from my hand to armpit—and the final wind-up in the Longview hospital with second-degree blood poison.

I thought of the second lancing of my hand with unsuccessful results, and two days later the feeling of horror that swept through me when the overworked young doctor announced in a hard, cold-blooded voice that amputation above the wrist was necessary, that it was mandatory that I get immediately in touch with my nearest of kin.

Like Simrose, I was violently ill at the time. It was five o'clock in the evening when the doctor announced his decision to amputate. The operation was scheduled for one o'clock in the morning. As he hurriedly left the ward,

I remember shouting at him that I refused the amputation, and that I would never give him the address of my family—that any decision that was to be made would come from me.

And after the doctor left and the minutes ticked relentlessly on towards the loss of my arm, I remembered sweating out the dark, helpless hours on that hospital cot.

I don't know how long I had been delirious when out of a distant haze and a fit of vomiting I saw the face of my old Los Angeles boyhood friend, a manager of his father's East Texas oil empire, twenty-four-year-old Gordon Green Guiberson.

I tried to grin up at Gordon. I can still hear those carefully-thought-out words of his. By the tone of his voice and the look in his eye, I caught the feeling that he had been trying to get through to me for some time.

"Listen, kid," he was saying. "Try to concentrate. Just for a few minutes. We're all set for action, but you've got to do your part. Do you hear me now? All right then, now I'm telling you for the third time, you're gonna lose your arm tonight and maybe your life here in this slaughter pen. Our one chance is Dallas. A train goes through here in fifty minutes headed for there, and you're gonna be on that train, see? I've had to work fast. We did everything we could to arrange your transfer from Longview to Dallas Hospital and the best surgeon in Texas, Doc Stone, but I can't make this arrangement stick with this dim-witted hospital bunch here. They say the time is too short and doctor's orders are that you can't be removed from here under any circumstances. It has something to do with workmen's compensation. They've got the law on their side."

Gordon cleared his throat.

"I called Dad up in Dallas. He's made an appointment with Doc Stone at the St. Paul Hospital to operate to save your arm instead of amputate, that is if we don't arrive there too late—and if it's humanly possible."

I was too weak to answer but my head was clearing, and I began to take in the proposition that Gordon was planning for me. It was about 105 degrees in the hospital room and 115 degrees on the streets of Longview. Gordon was sweating profusely. The water was trickling down his neck and off the end of his nose. Now his eyes twinkled and he let go his low, earthy guffaw.

"Boy," he said, "our friends, Buffalo Kennedy and Bunk Lock, are going to be here any minute now. They'll have a set of extra clothes with them to replace your outfit which the nurses must have hid some place. Vi is going to be in front of the hospital with the car running and when Buffalo, Bunk and I get you through this hospital mob—and I'm telling you now, nobody's going to stop us—we'll be down to the station in ten minutes. Vi and I are going with you."

Blond-headed giant, Buffalo (Gilbert) Kennedy, six-foot-three, two hundred and thirty pounds, was a trouble-shooter for a big oil company. So fabulous was this two-fisted, former Texas Ranger's reputation in those toughest towns of America that before most mob fights or saloon brawls went into action everyone chorused: "Which side is Buffalo Kennedy on?"

Bunk Lock was a hard-bitten, iron-muscled roughneck—

another formidable opponent if you happened to choose the wrong side of an argument.

When they arrived Buffalo and Bunk laughingly shed off a layer of their clothing, picked me up, twirled my one hundred and eighty pounds as if I were a babe in arms, as Gordon pulled my new clothes on.

I'll never forget the sudden, hushed silence in the passage way and outer offices as we marched through the hospital, Buffalo Kennedy in the lead, the carved handle of his long-barreled six-shooter slanting forward out of the top of his Levi pants—the slow, gleaming-eyed, questioning look that he flashed from one side of the motionless hospital crowd to the other—and behind him the hard-faced Gordon Guiberson and Bunk Lock carrying me along as if they were toting a child.

From the time we left my hospital ward to the moment when the boys lifted me into the car, where Gordon's wife sat at the wheel, I had the feeling that these Texas men had been hoping that somebody would make the grave mistake of interfering with our march towards the front door. But even the guards, whose business it was to stop any such law-defying shenanigans in the East Texas oil-field hospital, melted into the crowd of gaping-mouthed onlookers.

It is because of those three heady, tough, adventure-loving friends of mine that I still have a left hand that has come in handy on such things as shoveling manure and pitching hay.

I could still see Doctor Stone and his gauze-masked assistants around me in the operating room of that Dallas hospital—my arm strapped to the table. I remembered Doc Stone saying that everything was going to be fast and not too painful. I remembered watching the doctor at his work—the terrific relief I felt when his knife eased to the core of infection.

Several days later I remembered asking him why the Longview lancings had failed to eliminate the infection, and luckily his answers and his explanations came back to me, eight years later, in a candlelit shack in a frozen world three thousand miles north of Texas.

This time the actors had been switched and I was to be the doctor, and Simrose the patient. The great Texas doctor's words sprang to my mind:

"They didn't cut deep enough. I had to reach down through the minor flesh area to the seat of the infection, which was near the palm of your hand. Another few hours and it would have been too late."

And there it was—the answer. Doctor Stone's answer— *cut deep.*

That's what I would do. Cut deep to the core of infection, and Simrose's hand would be saved.

It was as if a great weight had been lifted from me. From that moment on to the end of the operation, I never once had a doubt but that I was doing the right thing and that the outcome would be successful.

I eased Simrose onto the bunk next to the table.

"Don't worry about strapping my arm down," he said. "I won't pull her back. Anything will be better than those tom-toms that're beating away in there now."

And Simrose was tough. He didn't flinch—my knife and plier-handled razor blade didn't miss the mark—and I didn't get squeamish when I found the core and cut in the drainage holes.

And then the dressings were on and it was all over—and Simrose was saying, "Wowie, wowie. I never felt more relief in my life."

And then the Bear was whining and complaining about my not putting the evening meal on the stove.

A week later Simrose drove his team and wide-racked hay sleigh out onto the meadow towards a distant haystack. He waved his pitchfork at me, and I raised my hand at him from my saddle horse as I trotted off in the direction of Nazko.

CHAPTER XIII

Social Life on the Nazko

THE DAY I STARTED out for Nazko from the Pan Meadow, the weather, which had milded-up after the Frozen Drive, took its usual turn for the worse. It was thirty-five below zero on Lashaway's thermometer when I rode in there. Lashaway insisted I spend the night with them and get an early start the next morning.

"You can make it to Albert Wilson's ranch in one day from here," said Lashaway. "About forty, maybe fifty miles. Too cold to sleep out in the bush now."

For once I used some common sense and let Lashaway talk me into it. Several of my toenails had turned black and were dropping out as a result of the Frozen Drive, and the soles of my feet were tender underneath the heavy growth of calluses. I had no desire to frost them again.

On this trip I rode Montana, and led a pack horse. Lashaway's fierce oversized mongrel dogs made a run for the Bear when they first saw him on our arrival the night before. But with his mane standing straight in the air he walked snootily and stiff-leggedly away from them, keeping close in under my saddle horse's heels. The Kluskus Indians will not tolerate their own dogs jumping a visiting fireman's dog. Lashaway's mongrels took a violent clubbing from several Indians who had witnessed our arrival. The Bear came out from under the feet of my saddle horse and stood by, showing visible approval, while I yelled at the Indians to let the dog pack go.

"They've done no crime," I hollered. "My dog can take care of himself. It's his own fault if these dogs work him over."

Of course the Bear took advantage of the situation, and from then on, if ever a Kluskus dog bared his teeth when he was around, the Bear looked tolerantly at the offender with a kind of white-man-supremacy air. His psychology certainly paid off, for, during the years in which a great deal of my time was spent with the Indians in their camps and villages, I never saw the Bear in an actual fight with the Indian dogs.

These Indian pack and hunting dogs are not Huskies, or Malemutes, or any particular breed at all, but they are bigger, fiercer-looking and far more dangerous than the mongrels you see around the garbage pails and empty lots of towns and cities.

A winter sleigh road connects Nazko with the Kluskus Indian village. Wagons or Bennet buggies can negotiate the rocks, boulders, muskegs and stumps for a few months of the year, but it's hard on equipment. However, as a saddle-horse trail it was a godsend to me while we were wintering cattle at the Pan Meadow, Lashaway's and Joe Spehar's at Nazko.

I rode into Albert Wilson's just before dark. Albert was a medium-sized, dark-haired fellow, about twenty-four years of age. He was batching it out on this lonely, stick ranch. I had met him several times before at Nazko and got the impression that he was lonely and not particularly happy about his isolation.

Here he was with no close neighbors or friends, and with the price of beef at only two cents a pound on the hoof, hardly affording him tobacco and beans money. He wasn't in a very cheery position.

Albert met me at the door of his cabin and went with me to the barn to feed the horses. We walked back to his house with the Bear leading the way. We were both talking at once. Albert opened the door, I turned around to emphasize some point I was putting across, stepped in ahead of him, and fell several feet into the cellar.

There was a ringing sound in my head.

Albert yelled at me from above, "I was trying to tell you to walk across this board, but you didn't hear me. Are you O.K.?"

"I'm fine," I said, crawling up out of a pile of potatoes. My shoulder felt as if it were out of joint.

"Climb up that ladder there in the corner," directed Wilson.

I found the ladder. It had three rungs. I had to stretch out a long way to make each step. I heard a slight, ripping sound in the seam of my pants. The top rung started to give way as a two-inch nail pulled out, but Albert grabbed me by the shoulder and I crawled safely out onto the cabin floor.

"I pulled out those rotten old floor boards in front of the door," explained Albert. "It was getting to be a regular trap there."

"It's lucky you pulled them out," I said. "Somebody might have got hurt."

"That's right," said Albert, as he poked a stick in what was left of a broken-down cookstove. "Joe Spehar was up, two weeks ago, looking for some cows of his. He stepped on the rotten board in the middle—just where you fell in—but I always step over that board. Knew it would bust through some day. I told old Joe to step over, but he didn't know which board I meant. He stepped right in the middle of it. The board broke and Joe landed down in the cellar in the spuds within three feet of where you did." Albert paused a moment while he struck a match and touched it to the kindling in the stove. "I guess the fall shook old Joe up pretty bad," continued Albert. "When I got him up here, he started pulling up the floor boards and smashing them with that axe there in the corner. So I helped him. He threw the boards out the door, said it would be safer when a person could see the open cellar. Better than stepping on a rotten board, and maybe getting hung up on his way to the basement—

"Here, here," said Albert, "we're talking too much. Let's have some coffee. Watch out now. Don't step on that board next to the bed. You're pretty heavy—whoops— O.K., that's right, walk on the middle board. Good going. Now sit on the bed. It's O.K. I'm going to buy some lumber next spring and put in a floor. Somebody's gonna get hurt in here yet."

"Well, I'll tell you, in these small cabins you don't need a great big cellar like this one," I told Albert, "just a small hole about four feet square. That's good enough for any man."

"Say, Rich, can you play any musical instruments? I've got a banjo, two guitars, and an accordian. We can play some pieces tonight."

"Where's that coffee?" I said.

"By God, that's right," said Albert. "Just a minute now. Whups—it takes so damn long to get anything done around here. A fellow's got to watch every move, or bang, he's down there with the spuds."

"Albert," I said. "Coffee."

I was sitting on the bed with the black pit immediately at my feet.

The cabin was about twenty feet long by twelve feet wide. The big opening started at the door and extended back some ten feet to the living quarters, where the stove, a sack of flour, a hind quarter of moose, two cats, several apple boxes nailed to the wall and an old round-up-wagon chuckbox took up the remaining space. A two-by-twelve-inch plank extended from the door across the big open cellar to the living quarters.

I was glad that I didn't walk in my sleep.

The following day Albert decided to throw in with me on the trip to Joe Spehar's.

"I've got to buy a clean shirt and some oatmeal and stuff," he said. "I might as well go down there now as next week."

We fed his stock, saddled up and prepared to leave.

"You better lock your cabin door," I said to Albert. "Some damn fool is liable to drop in to see how you're doing, fall into the cellar; maybe get killed or crippled up."

"No," said Albert. "We've never locked the cabin up.

Bush hospitality, you know. If anybody gets hurt in that house there it's his own fault, not mine."

We rode away with the Bear walking proudly in the lead. It was still kind of frosty, the mercury holding its own at around thirty below. I wanted to stop at Larkie's ranch, some twenty miles towards Nazko. I had never met Larkie or his Indian wife, but I had heard a lot about him. But Albert was not too enthusiastic about the idea.

Larkie was reputed to be a character—a big, cheerful bushman with a great sweep of mustache, an unusual sense of humor and not a worry in the world. It was commonly held that Larkie would be the last man to get high blood pressure or a heart attack from overwork or worry.

As we rode along, Albert told me about Larkie's latest money- and labor-saving scheme.

"There's a dangerous mud lake right near Larkie's," said Albert. "His cattle are always getting bogged down or drowned in it. Every year it's the same story. But Larkie told his woman that from now on any critter that kicked off in that lake would be stew and steaks for the family. He wouldn't lose anything that way.

"It's an idea all right, but things don't always work out. Larkie took a vacation trip to Prince George. Well, two big cows got out on that rotten ice on the lake a few days after he left. They drowned all right. Bloated up as big as elephants and floated to the surface. I saw them after they'd got froze into the ice.

"Well, sir, Larkie must have carried through his plan. I saw where somebody had cut the rotten old carcasses up with an axe. Took off front and hind quarters. Sure as hell Larkie is eating that meat right now.

"His woman went back to Kluskus to visit her relatives. She's not got back home yet. That was a long time ago. Guess she's waitin' for Larkie to finish off that meat before she'll come back."

By the time we reached Larkie's, my feet were tingling and I was ready to accept any cabin hospitality that was offered, but Albert said, "We better keep on riding. I never stay there myself."

Two-hundred-pound, heavy-built, Larkie met us at the door and boomed out his welcome. I immediately realized that at least we were not going to be in the presence of mediocrity. Larkie was an obvious extrovert.

"Come in, Albert," he hollered. "Too cold to ride on now. Lots of grub, lots of hay for the horses."

Albert pointed at me. "This is Rich Hobson. I'm riding on with him to Joe Spehar's. Sure gettin' cold, isn't it?"

"Rich Hobson," yelled Larkie. "Come in, Rich. I hear all about you a long time."

Now Larkie roared with laughter. "Ha ha haw." He grabbed his middle and doubled up. "Rich Hobson—haw haw."

I didn't quite grasp Larkie's humor, but I laughed with him. Albert remained silent.

Now Larkie roared forward towards me, Montana and my pack horse. I quickly slid out of the saddle and Larkie grabbed my hand. Montana stepped anxiously back a pace or two and I let my bridle reins go. Albert didn't get off his horse.

The Bear walked up to Larkie and began sniffing at his World War I surplus pants. Night was creeping down on us. I had a feeling that Albert and I were about to continue a long, cold, saddle-horse ride to Nazko in the dark.

"Look out for that dog," I snapped at Larkie. "He doesn't like anybody to shake my hand. You make one false move and he'll take a leg off you. Just move slow and careful."

Larkie became suddenly interested in his new-found dog friend. The Bear was waiting for me to give the word. The hair along his back was standing straight up. He licked at Larkie's army pants. Larkie let go my hand and reached down to pet the dog.

"Look out," I said. "Don't touch him until he knows you better. He's bushed, Larkie—bushed as hell. Been back there at Batnuni too long."

Larkie withdrew his hand. The Bear looked teasingly up at him.

"Fine dog," he said. "Great dog. Can see it as soon as you look at him. Good boy, dog—that a boy, nice dog."

"Rich and I have got to be in Nazko tonight," called Albert, from his horse.

"Come in, Albert, come in," bellowed Larkie. "You haven't stopped at my place for five years. Come in, Albert, come in."

Larkie's hospitality didn't seem to have any effect on Albert Wilson, his nearest neighbor. I could see that a decision had to be made—and quickly.

"O.K., Albert," I said, pointing up at him. "Crawl off that jughead. We're spending the night here."

Larkie roared with approval.

"This batching is tough stuff," I said to Larkie.

"Go to the barn," yelled Larkie. "Feed 'em the best hay." Larkie quit laughing for a moment and faced his reluctant neighbor. "You boys take care of the horses and I'll cook up the steaks."

Larkie turned towards his weather-beaten log cabin. The Bear watched him. Larkie opened the door. A strange smell reached my nostrils. The Bear walked nonchalantly towards the door, looking tolerantly back at Albert and me. Larkie turned. He looked at the Bear.

"Beef steaks for us all!"

The strange smell drifted strongly towards us. The Bear's tail was describing a wide arc. Larkie reached down. The Bear licked his hand approvingly. The door closed behind them.

"Let's get the hell out of here," Albert said.

"You're just getting cabin fever with your neighbor," I said. "You fellows have been seeing too much of each other."

"About twice a year," grunted Albert as he crawled down off his saddle horse. We led the cayuses into the barn.

"We should have rode on through," said Albert.

"To hell with it," I answered. "It's damn nice to have a roof over our heads, a good meal and a warm place to bed down in."

"God," said Albert. "If you only knew."

Larkie's log cabin was about twenty by fifteen feet inside. A large proportion of his floor was, like Albert's, gone. But there was no basement opening to worry about. We walked about on the two-by-six stringers.

"Got the 'flu," explained Larkie. "Ran out of wood. The wife was away—had to do something. There it was, starin' right up at me—the floor. I broke out the boards with the house axe. Had enough wood here for a week. Haw, haw.

"Going to Nazko next week. The old schoolhouse there. Nobody using it now. I'm gonna rip out the floor boards and bring 'em back here. All I have to do is to saw a foot off each one; pull the nails, lay the boards, and, presto! a new floor for Larkie; and no harm done."

Larkie patted the Bear's happy head. The dog had eaten so much his sides were swelled. He wagged his tail in marked approval.

It was getting hot in the cabin now. A bit too hot for Albert. He squatted on his heels near the door.

"I'm gonna open this door," he said.

Larkie was cooking the steaks. They certainly had a strange smell about them. He didn't hear Albert.

Larkie talked as he cooked.

"Nothing like cooking a piece of meat until it's cooked," he said. "Cook it right through. Trouble with these restaurants now-days—they don't cook their steaks. If you're gonna cook a steak, I say cook it not a half-hour, cook it an hour. Cook it for six hours if you've got the time. Keep turning it over. Keep lots of grease in the pan; but cook it. Then you get the flavor.

"We'll eat these steaks up first," said Larkie. "Then we'll have some more."

He stepped over to a forty-five-gallon barrel standing in the corner. I thought it was a water barrel. I was wrong. Larkie lifted off the lid. A horrible smell rose up. Albert faded hurriedly out the door as Larkie's arm came up and out of the barrel with a large chunk of pale, dripping meat.

I must have had a strong stomach in those days, stronger than Albert Wilson's. I just sat there on the sawed-off old log stool, and watched Larkie tear the meat in two, hand the smallest piece to the Bear; and, whipping out his hunting knife, cut the remaining piece up in hunks and throw them into the large iron cooking pan.

I said, "There's nothing too good for a cowboy, Larkie, even his own beef."

Larkie roared with laughter. He slapped his sides. "That's a good one," he bellowed. "Nothing too good for a cowboy."

I was able to eat two steaks. Larkie expected me to eat more. Albert retired to the barn where he spread out his sleeping bag.

"These young fellows can't take it," roared Larkie. "Trouble is they're all spoiled. Got to have everything dished up to 'em on a silver platter. None of 'em eats enough. I always say, just keep on eatin' lotsa good wholesome food, and you don't have to worry about anything. Just eat, I say."

The only open space on the floor where there was room enough to unroll my bed was next to Larkie's pole bunk. I didn't spend a very restful night. Larkie was a snoose chewer. He kept one corner of his mouth filled with the stuff even when he ate. He took a fresh can of snoose to bed with him, spat the old wad out of his mouth, and, with a generous refill, started to snore and chaw.

I had a canvas pack mantle over and under my bed, but part of the spray of Larkie's well-aimed wads kept making a splattering sound on the edge of my canvas cover.

There was no place else I could move my bed to, for there wasn't much floor space available.

The following day, as Albert and I rode into Joe Spehar's, we met Letcher Harrington. He yelled his greetings. I noticed at once that he had lost a lot of weight and he didn't look too well to me.

"Joe Spehar relented." Letcher grinned. "Bawled me out for not moving down to his cabin. But what's a man gonna do? I'd already got the gable ends of my cabin up; and I'll be damned if I'm gonna quit now."

The small log cabin was built, the barn nearly completed and the cattle in excellent shape. Letcher had turned in an almost superhuman amount of work.

Joe Spehar met Albert and me at his cabin door.

"That gotdamn Letcher," boomed Joe. "He's gonna kill himself working. I told him he could come down here and stay at the house. But no. Letcher is bullheaded. Gonna stay with it, he said. But if he keeps it up he's gonna get seek."

Joe was in a good humor.

"I'm getting even with that damn Paul Krestenuk," he said. "That big Rooshin stands with his dirty gumboots on my outhouse seat. Every time he goes by on his freighting trips. Now I've ripped out the boards and put in just one two-inch pole in the middle. Paul stand on that pole and he's gonna fall into the hole.

"Haw, haw. I set trap for Paul Krestenuk. He's gonna fall in sure. Then I go out there and laugh at Paul. 'That teach you to have better manners,' I'm gonna call down to him. 'Next time when you come to Joe Spehar's, you use your good manners.' "

Why Paul Krestenuk chose to stand on Joe's outhouse seat I never knew, but Joe claimed he never sat. The outhouse was a definite trap all right. Joe had sawed one end of the single pole almost in two, on the underneath side. Joe Spehar certainly had a great sense of humor.

Of all the characters I have thrown in with in this north country, Joe Spehar definitely deserves a place well up towards the top of the list. When you arrive at Joe's dirt-floored shack he expects you to stay on indefinitely. You could stay for a year if you wished, as long as you didn't bother Joe with small talk or any kind of talk, and ate his specially constructed mulligans.

I must say that if you didn't object to a fair amount of garden silt in your food, Joe's mulligans were hard to beat. Joe could raise a garden where nobody else could. He had that green-finger touch. His mulligans were made in a monstrous tub with vegetables as the base. Joe just peeled the outside leaves off of two seven- or eight-pound cabbages and dropped them in boiling water. A hatful of carrots of proportionately enormous size were hurriedly washed and dumped in. Beets, spinach, heads of lettuce, parsnips, horse radishes, handfuls of home-grown garlic bulbs were added. And then, last but not least, several smoked rainbow trout were entered, heads and all.

This fantastic conglomeration of everything, including the dirt, actually mellowed itself into a delicious dish. Joe cooked his mulligan every six days. The first day Joe attacked the thing with a reasonable amount of self-control and discretion. The following day he invariably let himself go. For his size I'll make a bet that when

Joe Spehar is in his best form he can eat more mulligan than any man, pound for pound, in the world. I am proud of my own vast eating capacity, but when it comes to mulligan, Joe Spehar can eat rings around me.

At the end of the second day of his mulligan orgy, Joe groans, falls back on his pole bunk and, holding his stomach and rolling from side to side, proclaims in his low, guttural voice, "I'm seek—I'm never gonna eat again."

Joe lives up to his resolve for the next four days, coffee and canned milk being his only form of nourishment. On the fourth day, as regularly as clockwork, Joe starts to complain about being hungry.

"I'm hungry," bawls Joe, his deep voice rumbling like a starving bull. "I'm hungry. It's got to be mulligan. I'm gonna cook up a big mulligan. That's what's good for me, a great beeg vegetable mulligan with feesh in it. That's what's a matter with ceevilization today. Young fellows don't eat the vegetable mulligan. That's why they can't work. Young fellows got no bottom nowadays. They gotta eat all thees fancy grub like butter and jam and seerup.

"I was born in a cave. Lived on vegetable mulligan. Never got seek—"

By this time, Joe would be drooling at the mouth, his fish and vegetables lined out on the big cabin table like an assembly line, and the huge washtub boiler bubbling away.

Albert and I had just completed a mulligan meal with Joe, on this trip to Nazko, when a loud bark from the Bear brought us all hurriedly to the battered door.

Alec Harrington was astraddle his buckskin cutting horse, dangerously close to the caved-in end of the porch.

"Joe," snapped Alec, "have you still got that ten-gallon keg of parsnip Tiger Soup cached away? We're putting on an old-timers' reunion tomorrow night at my place. I'll make it good on the wine."

"It's time for a party," he roared. "I'm seek now. Too much mulligan—but tomorrow we'll bring the Tiger Soup and come down and tear up your outhouse."

Alec said, "O.K. Bring Rich and Albert and be down to my place by dark."

Alec swung his buckskin about and trotted off into the jackpines.

Albert watched Alec disappear and turned to me.

"About every two years, the Harringtons throw a real old-time cattlemen's brawl," he said. "You'll see a real explosion—a Nazko special."

Nazko pioneers, the Harrington boys hailed from an old Virginia family. Cheerful, middle-aged Flem and Alec Harrington had drifted north into the Chilcotin from their home State. Their wide experience with hot-blooded saddle and carriage horses in the South stood them in good stead on the ranges of British Columbia's interior, where the horse plays such an important role.

While Flem plunged north into the sparsely settled country of the Nazko River Valley, Alec held down the foreman's job on a big Chilcotin ranch. Later he followed his older brother into the Nazko, and there they both pioneered cattle ranches.

Several years after Flem's and Alec's arrival in Nazko, their young brother, Letcher, also caught the call of the wild and had arrived there, with his wife, to dig into a stretch of open land that lay between his brothers' spreads.

Fred Rudin, another Nazko pioneer, who tired of the high Alpine meadows of his native land of Switzerland, was at Alec Harrington's when we arrived the next night. For a pittance, not even worth mentioning, Rudin had taken on the government mail contract. Once a month, for twelve months a year, good weather and bad, fifty-eight-year-old Fred Rudin slugged the mail sack through the one-hundred-thirty-mile round-trip of mud, snow, ice and flood, Nazko to Quesnel and back. His was one of the longest, and, without doubt, toughest, wagon- and pack-horse mail routes on the continent.

Zalowie George, small, neatly dressed, well-read, one of the few Indians in the country who had taken out citizenship papers, had given up all Indian reservation and treaty rights, and started a cattle ranch of his own, was one of the Harrington guests.

This was the first time that I had been in the same gathering with Paul Krestenuk, legendary Russian frontiersman, the bull of a man who had slashed out the one hundred seventy-five miles of wagon trail from Nazko to Ulgatcho Indian village. Paul was only five feet ten, but he must have tipped the scales at well over two hundred pounds, all sinew and muscle. He had jet-black hair that grew well down on his forehead.

Krestenuk never needed a town haircut or shave. He furnished a steady-handed Nazko Indian with a pair of clippers, a straight-edged razor, long-bladed scissors and shaving cream, and, in return for this initial capital investment, the Indian took over the duties of special barber to Paul Krestenuk, a job he did to Paul's liking. Once a week, winter and summer, the Indian clipped Paul's hair almost to the bone, a little closer to the skull than a crew hair-cut, and gave him a shave every second day.

Paul is a striking-looking man, his well-rounded head is supported by a bull neck, and his eyes are the size of apricots.

Krestenuk's main center of operations for his far-flung trading posts and fur-buying expeditions was his all-around, every-item Nazko store located on the edge of the Indian village, three miles from Alec Harrington's, where this old-timers' reunion was taking place. Paul was not only one of the toughest and hardest-working men on the frontier, but also a dry-humored optimist. Gathered together in his two hundred pounds of close-knit mental and physical being, Paul had about every requirement that was needed to make a success on any frontier.

The din in Alec Harrington's ranchhouse was rising to an unprecedented high when the Lavington brothers rode their saddle horses up the steps and onto the ranchhouse veranda.

Dude and Art Lavington, both top cowhands who had followed cattle herds, chuckwagons and horse remudas from the time they could crawl a horse, from the prairies and foothills of Alberta, and the grassy park lands of Montana, to the wolf-haunted jungles of central British Columbia, had finally decided to throw their ropes down in the Nazko country. These open-minded, hard-hitting, but easygoing cowboys were a great addition to our sparsely settled back country and every one of us had been working on them to choose this section of the land to drive in their stakes.

I didn't know it at the time but the Lavington boys had made their decision, and not so long after this get-together at Harrington's, Art bought out the Harrington ranches and Dude settled on Baker Creek, a few miles east of Joe Spehar's.

Now both boys got into the party in a hurry. The gang of us piled out onto the veranda in a large, whooping, close-knit group, shouting and waving our cups and glasses of Tiger Soup at the boys and their bewildered mounts. Art's horse shied backwards, hit a column holding up the roof—there was a splintering crash as the floor gave way under the combined weight of horses and men—and we found ourselves in a tangled, struggling mass on the hard ground, nearly three feet below the porch floor, where we had been standing happily waving our glasses a moment before. Now we thrashed about in the broken boards, logs, nails and splinters.

Most of the men struggled clear of the debris before the next catastrophe. The porch roof seemed to weave uncertainly for a breath-taking moment, and then with a high, grinding squeak, settled slowly down on top of Paul Krestenuk, Joe Spehar, myself and on Dude Lavington's saddle horse, the only ones who hadn't got clear of the wreckage fast enough.

I can still hear Paul's bellowing laughter, Joe's roaring, bull-like blasts, the horse's grunts and my own nervous laugh as the surviving members of the Nazko club staggered about trying to extricate us, their less fortunate brothers, from the ruins.

This slight incident in the Harrington brothers' old-timers' reunion merely added to the gaiety of the party and gave further incentive to the consumption of Joe Spehar's Tiger Soup. I heard later that Art Lavington subtracted the overall cost of a ten-gallon keg of Tiger Soup from the purchase price of the Harrington ranches, insisting that the clause in the contract, "ranchhouse intact," included the front porch. Whereas Alec Harrington claimed this was unfair, considering the fact that a Lavington horse had knocked the support out from under the roof.

CHAPTER XIV

Hidden Tribe of the Fawnies

JOHN JIMMY JOHN, his wife, Ellie, and four children lived with Jimmy's in-laws, the Missue tribe, on a lake called Tatelkuz, that nestled between the Nechako and Fawnie mountain ranges, some forty miles west of Batnuni Ranch.

For several years Jimmy John had been relaying to me the venerable old Missue's invitation to visit them. It wasn't until Jimmy left the Batnuni to return to his home in December of 1940 that I was able to take the time off to accept the invitation.

Before he rode out of Batnuni Jimmy said, "Best thing you can do is to follow my horse tracks over the Nechako Mountains to Tatelkuz Lake. I'll blaze trees once in a while in case it snows.

"You follow left fork of Taiu Crik till you get to a long, narrow lake running east and west. Follow down the north side of lake to the end of a muskeg and cross a beaver dam there. Don't cross any place but on that beaver dam.

"Ride ten miles down the edge of the muskeg, then you can look down at Tatelkuz Lake. We're on the northeast end of it. If you're not there the day before Christmas I come out and look for you."

A few days later, carrying in my pocket a small map drawn for me by Jimmy, and following in his still-visible tracks, I swung off the main Batnuni at the forks, and rode up Taiu Crik through the jackpine foothills of the Nechako Mountains.

The day was warm and the light crust on the foot-deep snow at Batnuni was thawing out. I whistled and sang as I rode along. Well up in the mountains I saw a moose browsing on the branches of swamp birch and willows.

A well-traveled, deep-rutted trail crossed Jimmy's horse's tracks, and I didn't have to look very closely to see that it was made by a pack of northern wolves. This wide, hard-packed trail zigged along in the same direction in which I was traveling. The band was covering what I knew to be a huge circle through game country, and within three weeks these killers would be covering the same territory again.

In places the tracks fanned out in the bush, and I knew that the outside wolf scouts had winded game. I counted twenty-two single tracks in this pack; some of them were big—the size of a small plate.

Once the Spider snorted and jumped to one side. He refused to follow in Jimmy's horse's tracks. I swung him in a half-circle around the spot, and wasn't surprised to see blood, hair, and hoofs of two moose, lying in a messed-up circle a few yards from where I rode. From the size of the teeth-scarred hoofs, I figured one animal had been a cow, and the other a big calf. The old girl had fought for her calf to the last, but she didn't have a chance—no more of a show than any other cow or bull moose, deer or caribou, when fate throws the wind from game to wolf.

And so I rode on, my spirits not quite so high.

It was four o'clock in the afternoon and just getting dark, when I reached the edge of the muskeg. I was looking for the beaver dam Jimmy John had told me of when I suddenly heard wind roaring through the jackpines. A few minutes later I stared at a great white cloud moving over a ridge ahead. In a matter of seconds a blinding sheet of snow swept down upon us.

All objects at more than twenty feet suddenly blacked out. The blinding gust of snow struck me full in the face, and the Spider lowered his head, arched his neck, and plunged ahead. I ducked and pulled up my moose-hide collar.

The bite of the fine snow with the terrific wind behind it was hard to take—and then before I knew it we were traveling with the storm at our backs. I realized—none too quickly—that we had swung about, and were traveling east instead of west, but to add to my confusion the wind behind the fine, driving snow made a great swirl around us, and then came in hard from the opposite direction.

This time there was no change—the blizzard steadied itself.

The Spider fought head on into it; and then veered sideways. It was no use. No man or animal could face that wind.

For a moment I was completely bewildered, and couldn't figure out which direction the storm was coming from. We were in a high gap in the Nechako Range. When the storm first struck down here, the strange air currents in the pocket tossed it about, but when the full force of the blizzard had gathered behind its front edge, the wind had settled down in its natural course.

There was now little doubt but that we were in the sinister grip of a northeastern Arctic blizzard.

The real Arctic blizzard seldom strikes this country, but when it does the mercury plunges to thirty or forty degrees below zero, with a northeast wind blowing a smothering mass of fine snow forty to sixty miles an hour, parallel to the ground.

These savage storms are the ones you read about, where men get lost between house and barn, are frozen stiff beneath a snowdrift within an hour. Many wild horses perish on open ranges, and a terrible death toll is taken of wild game.

I realized there wasn't a second to waste, for the driving, biting cold was already creeping through my clothing. The Spider had swung about again and was fighting his head to run with the storm. I crawled stiffly down out of the saddle and began to grope about for trees. It was hard to keep my balance, plowing across uneven ground covered with snow that came above the knee. Then the prickly branches of a wide spruce tree hit me in the face.

I tied the two horses to a small, green jackpine in the lee of the spruce tree. Feeling around in the dark with half-frozen hands, I managed to untie the ropes and get Nimpo's pack off. The wind and driving snow roared and wailed around us.

Now to get a fire started. My nose and lips were getting numb, and so were my feet and hands. It was so dark, and the snow blowing so thick that I had to grope about for dry wood. I didn't dare move ten feet away from the tree. If I did there was a big chance I wouldn't find it or my horses again. I stumbled over a heavy top of a jack-pine sticking out of the snow and pulled it out; but it was caked with ice; too soggy to start a fire with.

I hadn't brought along my usual jackpine shavings. The situation was grim. The cold was unbelievable. I scooped up a handful of snow and rubbed my mouth, cheeks, and nose until I could feel a prickly sensation. Man, how those parts hurt when the circulation started up. I just had to find some wood and get a fire started—and quickly.

My leather mitts were frozen stiff. I had trouble bending my hands. Here was the big danger—when a man's hands freeze he is then completely helpless.

I thought about the lower limbs of the spruce. Nine times out of ten the few bottom, inside branches of a large spruce tree are dead ones. Ducking in under the branches of the tree, I felt above my head. There were dead limbs in that tangle. I had trouble bending my fingers around the handle of the saddle axe; I started hacking away at the limbs above my head.

"What luck!" I said to myself. "A pile of good brittle ones. I've got to hurry—can't feel my feet any more."

There was a small, bare, spruce-needled spot on the lee side of the tree. I squatted down—slashed dead branches into small slivers and sticks. The wind swished and swirled around the tree. It didn't hurt my face and lips any more—I realized they were frostbitten.

The horses whinnied and thrashed about. I took off my right mitt and reached in my pocket for my jackknife.

I thought: These next few minutes are going to tell the tale—if I can't get a blaze going in ten minutes my hands will be frozen solid.

I couldn't thumb open the blade with my fingers. I bit it out with my front teeth. Then I slashed off some thin shavings, and my fingers cramped in a half-closed position; I broke off a limb with a few dead spruce needles clinging to it and made a little pile of the dry stuff.

I got out some matches. The heads flew off—they were wet. Now for my Kachie bullet lighter. It was in the bottom of my pants pocket. I pulled off the top with my teeth. I remembered one of those radio competitions where some man ran his bet up to a very high figure and the announcer said:

"Now light your lighter. If it lights the first time you win double the amount—if it doesn't you lose. Do you want to try it?"

I could still hear the cheers of the audience and the announcer bellowing into the microphone, "It lit the first time! Hurray, Mr. Splotz, you win the money."

Now I thought: Here is more of a gamble than Mr. Splotz's. If this lighter doesn't work now—well— what the hell!

I couldn't bend my thumb over the wheel. My right hand was only capable of holding the lighter. I scraped it hard against my pants leg. The wheel spun; a thin light reached out—but the wind snuffed it out.

I started scraping the wheel of the lighter again. A tiny spark took hold of the wick—I jerked off my hat with my stiff left mitt and shielded the glow from the wind. The flame grew. It flickered. I lowered hat and light to the little pile of shavings.

A fierce blast of wind swirled through the hat and the light half died, but caught hold again. A threadlike line of blue smoke rose slowly above the tiny spruce flame, then flattened and swished away with the wind—but the point of light grew and then sputtered and cracked through the spruce needles.

I added more small dry limbs to the fire—and then all at once the flames roared. The strange little world around the big spruce tree suddenly opened up. Survival—life—warmth reached out at me.

I could see the horses—and close to them a dead jack-pine tree. I plunged through the snow and whacked it down. I cut more limbs. Split a small, green spruce tree into eight-foot slabs, and crowded the chunks on both sides of the fire.

Luckily I had packed a small ration of oats for the horses—enough for two feeds. It was the tail end of the one sack of feed oats I had packed from Quesnel to Batnuni the previous fall. I poured the works onto a canvas tarp and spread it between Nimpo and Spider. I rubbed my face, hands and feet into painful feeling again, knowing better than to crowd in close to the fire too quickly.

I melted snow water in a lard pail. I don't think hot coffee ever hit the spot more than it did on that night of the Arctic blizzard. Outside the tiny world around the spruce tree, the blizzard raged and pounded and screamed across the land. But by the grace of God my horses and I had it whipped.

The dark, blizzard hours crept slowly by. Night seemed endless. The light of the fire revealed great snowdrifts piling up beyond the spruce.

I'll have to stay here until the blizzard quits, I thought, and then I'm going to have a tough push to get through to Tatelkuz Lake.

The vague light of a new dawn broke through on a spruce-and-jackpine world buried beneath great slopes and sharp-edged banks of drifted snow. With the approach of day the blizzard gave a few final swirls, and moaned away through the distant treetops, leaving behind it an ominous silence, as relentless in its deathly cold calm as the blizzard had been in its savage fury. But I was lucky. Usually an Arctic blizzard lasts from two to five days.

That was a long, hard day on the horses. Not every cayuse could have plunged through the drifts for fifteen miles to Tatelkuz Lake; but the Spider and Nimpo were super-horses—short-legged, and built to order for this type of drift-plunging. They hit into soft drifts to their shoulders, and plowed and tunneled their way through them with incredible energy and a certain graceful ease.

Late in the afternoon we emerged from the green, snow-smothered forest on the edge of a wide, slanting burn, and for the first time I looked down on the long, white, egg-shaped body of Tatelkuz Lake.

A tiny cluster of cabins, with a ruler-straight line of blue smoke rising from one, dotted a slope on the lower end of the distant lake, and I swung the Spider in a direct line for that happy-looking spot.

I noticed several horses tied up at a hitching rail, and others nibbling at hay chaff in a corral. The buildings consisted of one large, square log cabin, and four or five smaller ones. All of them were perched on a high, open knoll overlooking the great expanse of lake, and a flat prairie at its lower end.

A big creek flowed out of the lake below the cabins, and I was surprised to see that it was still open and running. The prairie rolled away to the north where it slid behind a growth of bullpine and giant poplar trees.

Out a short distance from shore, several Indian children, and an old beshawled Indian woman fished through holes in the ice. I saw Jimmy John down by the corrals throwing hay to a bunch of horses. I rode down the hill and waved. He greeted me with a big grin.

"Good boy, Rich," he said. "You made it through the storm. I knew when that cold blizzard broke that you were on the trail. That cold, he always know when you gonna go some place. Come on to the barn. Those cayuses need lots of good hay."

I snapped stiff-leggedly off the Spider, who jumped just as quickly to the side letting fly a loud snort. I hung on to the bridle reins and we led both horses towards an old-time log barn with a partly collapsed roof.

"Sure tough, that Spider horse," said Jimmy. "Not played out yet. I was gonna ride out tomorrow and track you down," he said. "That was sure bad storm."

We took care of the horses and started up the hill towards the big, square log house. Jimmy's two older boys, Alfred and Joseph, ages eight and nine, ran towards us. They each dragged a gunny sack.

"Look!" yelled Alfred. He dumped the contents of the gunny sack out on the ground at our feet.

Three large, white-colored, snake-shaped creatures, weighing perhaps five pounds apiece, wiggled in the snow.

"Me," yelled Joseph. He also produced several of these strange, eel-like fish.

"Ling fish," said Jimmy. "That's the kind of fish some white man call fresh-water cod. You'll see. Best eating fish. Boys catch lots. Tonight we eat ling fish."

The boys gathered up their catch and ran up the hill ahead of us.

Several girls and women grinned at us from the long bench in front of the house. I recognized Jimmy's wife, Ellie, at once. She smiled at me and nodded her head.

"Hello, Rich," she said in a low, soft voice. "We all worry about you in that storm."

"Hi, Ellie," I said. "I ride a tough horse."

Ellie was a very good-looking, shapely, neatly dressed Indian woman with about five sisters. All of them lived here at old Missue's, their father's, village. Besides Jimmy and old Missue the only other grown man in the village was the wild and fierce-looking Cultus Johnny, a fabulous character who was married to one of Ellie's sisters. Jimmy John had always complained to me about Cultus Johnny. Cultus means "bad" or "ornery" in the Indian language.

Jimmy had said, "That damn Cultus—he sleeps all day in one corner of Missue's house. Everybody he bring Cultus food. Cultus never works. Just lay there on his blanket and laugh and snore. We all divide up, potlatch grub and meat. And Cultus he eat up everything. Never bring in any grub himself. Just lay on that blanket and laugh and sleep and eat and snore.

"I don't know what's a matter that Cultus. Don't know why any woman crazy enough to marry that kind of man. I don't know why I gotta work hard all year and feed that skookum ugly face for nothing."

"What about old Missue?" I asked Jimmy. "What does he think about Cultus laying around?"

"Once in a while Missue tells him to get out and work," explained Jimmy. "Haytime, or traptime in spring. But the old man he never bothers Cultus. He's blind—can't see what Cultus looks like so he don't mind if Cultus lay there on the floor."

"What's Cultus' excuse?" I asked Jimmy. "What does he say is the matter with himself?"

Jimmy grinned. "Cultus always say he's resting up. Still tired from the time he ran afoot from here to Nazko to show his girl he was best all-round man in this country. About ninety miles to Nazko. Cultus used to do a lot of foot running. On his own feet he could play out any horse in the country. He made it to Nazko without stopping once.

"Big Francis and Long John both ride saddle horse to Nazko the same time. But Cultus he beat 'em there by nearly half a day.

"After that Ellie's sister she admit Cultus Johnny is best man in country, so she marry him. Cultus never did anything since. Just lay there on that floor and laugh and eat my grub and snore all the time.

"Old Missue wants to meet you," Jimmy said. "Come on in."

The women and girls stepped aside. Jimmy entered the house and I followed.

It was as if I had plunged from a light and sunny world into a black dungeon. For a moment I couldn't see ten feet in front of me, then my eyes adjusted themselves to the dark room. The entire house was one big, bare room, about twenty-five feet square. A dilapidated heater and cookstove combination with a drum oven attached to the stovepipe sat solidly in the middle of the floor.

I was amazed to see that the only furniture, outside of the stove, was one large dresser with a flour-barrel attachment that glared at me from a spot close to the door. A pile of plates and half-opened supply boxes crowded against the dresser. The rest of the wall space was taken over by an array of assorted mattresses, blankets, canvas tarps and gunny sacks.

Jimmy said, "Come over here and meet Missue."

I could see well enough now in the dark room. I picked my way carefully through the pile of blankets and mattresses behind Jimmy towards the quiet figure of an expressionless old man who was rubbing a motheaten mustache with the index finger of his right hand. I could see at once that the venerable old Chief Missue was totally blind.

We reached the side of the old man and Jimmy and I stood there looking down at his still figure.

"Missue," said Jimmy. "This is Rich Hobson. He come long way to see you."

"Hello, Missue," I said.

The old man was squatting on the floor, his back to the wall. He was partially covered with blankets. He made no reply and his intelligent but impassive face did not change.

His only movement—the finger that rubbed his barely visible mustache—moved slightly faster.

Jimmy and I stood there in complete and awkward silence for fully two minutes. I began to think that Missue hadn't heard Jimmy's introduction, but knew better than to try to strike up a conversation. Missue, the master of the gigantic, undeveloped empire that spread from the Nechako Mountains across the distant summits of the Fawnies, was merely sizing me up in his amazingly clairvoyant mind.

Suddenly he grunted, then slowly began nodding his head. Then he stuck his hand out in my direction. He grunted a few Indian words.

Jimmy said, "Bend over. Missue wants to feel your face."

I dropped quickly to my knees and bent my head forward towards Missue's outstretched, wrinkled old hand. Missue ran his hand over my face and then my neck and shoulder.

"Good boy," he said in broken English. "Now I know you."

Jimmy gave me the wink and I straightened up.

Missue remained silent for a moment, rubbing his mustache. "You all same as Indian."

"Thanks, Missue," I said.

Long, comfortable snores had been drifting across the room from a blanket-strewn corner near the door. I had been conscious of these well-measured, monotonously regular sounds since entering the room. Each long, plaintive snore ended with a high, wheezing note. There was no variation between snores. Every one rose into the darkened stillness of the room with musical regularity.

Somehow I could understand Missue's reason for enjoying the company of Cultus Johnny—for the pleasant rising and falling of Cultus' music added to the feeling of serenity and well-being that hovered there in that dark, silent room.

As we reached the door Jimmy pointed down at the ragged but relaxed body that sprawled across an old, black-colored, army mattress and a pile of dirty army blankets and gunny sacks.

"Cultus," said Jimmy. "He'll keep on snoring until his woman wakes him up to eat. Then he fall back and start to snore again. Why do I have to feed his ugly face? I don't know why. Lucky Missue got lots of fish in that lake. Cultus eats two, maybe three, fish every time somebody wakes him up."

Jimmy opened the door and we stepped outside. The glare of the evening light struck me so hard that I had to shut my eyes for a moment.

The Missues were a branch of the Kluskus Indians. They were a kindly and benevolent people, isolated from the rest of the world, and even their own tribesmen, in a kind of Shangri-la of their own making.

It was easy to see that this family was a poverty-stricken one. But what these people lacked in worldly goods was made up a thousandfold in their unswerving faith in the Great Spirit.

Through the years I grew to know the Missue tribe well and many is the time that I routed my trips from Batnuni to Anahim or the Home Ranch through the jungles and muskegs of the high Nechako Mountains so that I could stop at Missue's.

There I seemed to undergo some kind of spiritual rejuvenation. I firmly believe that the combined thoughts of faith and goodwill that oozed out into the atmosphere from those good people were picked up by my sometimes sagging mind.

So strong was the faith of the poor and often hungry Missues that they actually believed themselves to be rich and well off in every sense of the word. Everyone—women, children, Cultus Johnny and even the large tribe of assorted dogs—was smiling, assured and happy, with the one exception of the more worldly and practical Chilcotin Indian, John Jimmy John, who, like the white man, occasionally worried about where the next meal would come from and how the bills would be paid.

Cultus Johnny, without doubt the ugliest and laziest man in the world, gave out a feeling of strength, security and goodwill. His hissing laugh was infectious. He had not a care in the world, for the Great Spirit (Jimmy John claims that this Spirit was actually Jimmy John) would always look after him when he was cold or hungry.

I consider that I was greatly privileged to have been accepted by the Missue Indians and taken as a brother into the tribe.

Many years later the Missues added Albin Simrose to their exclusive group—and Jimmy John, Simrose and myself formed a ranching partnership to develop Missue's great, but unproductive, black-soil prairie of Tatelkuz Lake into an alfalfa proposition that if fate hadn't intervened would have been a feed-producing stock ranch out of this world. But—whoa there, we're getting ahead of our story.

CHAPTER XV

―――♦―――

The Stampede Road

WHILE THE CATTLE FRONTIER north of the Itcha Mountains was being pioneered by men on saddle horses, and freight was landed in this remote area by pack horses and wagons, a visionary and energetic fellow named Stanley Dowling had been pioneering the first and only trucking business into Anahim Lake.

Stanley was solidly put-together, a compact one hundred seventy-five-pound chunk of muscle, raised in the city of Vancouver. For a while he followed the fight game and was a good, middle-of-the-road, middle-weight boxer.

Disliking the city, he tried farming on a small scale in the Bella Coola Valley, ninety-five miles west of Anahim. Stanley wasn't cut out to be a farmer. He found this out after a lot of hard work and the loss of his small capital investment.

Then for several years he worked for the Andy Christenson and Lester Dorsey cattle ranches at Anahim Lake.

It was while he was helping move pack trains from the coast and wagons from Williams Lake into this isolated area that he conceived the idea of improving the bush road in the worst spots and trucking in much-needed supplies.

I remember Stanley explaining his plan to me in front of the Christenson ranchhouse at Anahim Lake.

"Rich," he said, "there is a big opening here at Anahim. I figure a guy can take out fur, cattle and horses in return for goods trucked in from the wholesalers in Vancouver. I can haul out and sell the cattle and horses when I leave for town empty. I have sixty dollars wages plus fifty-five dollars for the two horses I just sold you. That one hundred and fifteen dollars is going to see me to Vancouver. I've got a scheme for financing my first load."

I said, "With one hundred fifteen dollars as your total working capital, Stanley, I would say you certainly need some kind of a scheme—and a good one at that."

Two months later, when Pan rode into Anahim, he found that Stanley's scheme was working out.

There Stanley was, at the end of the road, with a well-running, second-hand Ford truck, and three tons (equal to thirty pack-horse loads) of assorted paraphernalia, equipment, tools, machinery, dry goods, groceries and even fresh fruit and green vegetables. Stanley had forged into Anahim on the first truck into the country, with a good three thousand dollars' worth of outfit. It was hard to believe that only two months before he had slugged southward on his saddle horse, with only one hundred fifteen dollars in his pocket.

Stanley's scheme could be partly explained by a Ford advertisement in a Vancouver paper and several trade magazines. Stanley showed us the advertisements. The ad contained, in addition to a glowing description of Ford trucks, a small picture of Stanley in his Stetson hat, riding boots and chaps. The caption explained that cowboy Stanley Dowling was to drive the first truck ever to attempt the two hundred twenty-five-mile lug through mountains and mudholes from Williams Lake to the distant Anahim Lake country, where he was to deliver much-needed food and supplies to the isolated settlers.

This daring feat, naturally, was to be attempted with a Ford truck.

Stanley had a winch on his truck, logging chains, ropes, and every kind of jack and tool to help him make the journey. He told us he had used the winch often, that there were times when he was hopelessly stuck, that he had walked miles to ranches where he had hired four-horse teams, which pulled the truck out of the mud.

A short time after his arrival at Anahim, Stanley had completely sold out his goods. That same winter he trucked his second load of over four tons as far as Tatla Lake, where he wintered his truck, and, with teams and sleighs, freighted the food and equipment the seventy-odd miles into Anahim to his newly finished, twenty-by-twenty-foot log-cabin store and living quarters.

He paid for this building with twenty-five dollars' worth of provisions, a second-hand cooking stove, some needles and thread, three sets of horseshoes, and several pairs of moccasin rubbers.

During the winter Stanley really cleaned up. He had studied the fur business in his spare moments, and, in trading food and equipment to Indians for beaver, rats, squirrels, marten, mink, otter and fisher, Stanley proved he was not only a lucky trader but a cagey one.

His success was fast and fabulous. Two years later he drove a new Ford truck into his garage at Anahim Lake. Climbing out of the cab, he walked into his new store—a log building fifty feet long by twenty-eight feet wide. On its many counters and shelves, and in a warehouse, were stored a valuable stock of hardware, haying machinery, groceries, dry goods and drugs. He even had plans under way for one of the fastest and most unique stampedes to be held in British Columbia, where some of the toughest bucking horses in this part of the world were to unlimber.

Stanley hired two men to run his store, his post-office, and his small hotel, but he stayed with his one truck, drove it himself, traded it in for a new one each year, was constantly on the road and at the buying markets and wholesalers.

At the present time Stanley owns one of the smartest cattle ranches in west Chilcotin and a grain finishing farm near Vancouver; and Anahim Lake has benefited by his hard and heady work. A good road soon followed on Stanley's wheel tracks and the long, frustrating, and expensive, pack-horse and wagon freight has been replaced by supply trucks delivering much-needed and far less expensive food and equipment, including some of the luxuries that other people in the outside world take for granted.

After Stanley had established his store, warehouse and post-office, Pan said to me, "Friend, there's no gettin' around it. Civilization has crawled right up on top of us.

There's no use to go around the country tearin' our hair out over it.

"Our only way out of the mess is to adjust ourselves to it. Just think—we're only fifty miles from a real first-class store and P.O.—fifty miles over the Itchas.

"The first thing we got to do is build a road over those mountains to connect with Stanley's layout."

"Fifty miles of road is a lot of road," I told the Top Hand. "Particularly over a seven-thousand-foot mountain range of rock and muskeg and jungle."

Pan leaned confidentially towards me, and the long ash on his cigarette suddenly broke loose on my shirt front, missing my open collar by inches. "Watch that stuff, you imbecile!" I roared at the Top Hand. "You're not accomplishing anything by dropping dirty cigarette ashes over everybody."

"There'll be nothin' to buildin' that road at all." Pan grinned. "I've got an idea how it could be did. I'm gonna ride over the pinnacle and see Stanley."

Pan and Stanley Dowling had a business conference at Anahim Lake the year of the first stampede. What those two boys figured out changed a lot of things north of the Itchas.

At the business meeting Pan told Stanley what was on his mind.

"Now here's what I'm drivin' at," said Pan. "We want to get that fifty miles of wagon road built early this summer, and we ain't got the manpower or the money to put her across. Well now, Stan—we'll start from there.

"You need a lot of customers here to buy your

stuff—and you want to buy a pile of fur next winter and trade the Indians grub and ictus for same. O.K. Your biggest future fur country lies north of the Itchas—Ulgatcho trappers—Kluskus trappers. If a wagon road was slashed through from the Blackwater to Stanley Dowling's Emporium, Mr. Dowling would soon have a line of sleighs or wagons full of fur heading over those mountains."

"I see what you mean," said Stanley thoughtfully. "It would be the makings of my business here and would cut your Home Ranch's freight costs down to a minimum. I haven't got much ready cash lined up but it's a go. Have you any ideas how we can line up the workmen and the money?"

Stanley looked sharply at the Top Hand. He had already found out that Pan did a lot of thinking before he tackled a tough deal, and usually was able to scheme out some fantastic and yet simple plan of action.

"It's easy." Pan grinned. "Nothin' to it, boy. How about a tailormade coffin nail?"

Pan held out his hand. Stanley got up from the small card table in the kitchen end of the store and brought down a carton of Players cigarettes from a shelf. He tossed the full carton to the Top Hand who caught it as if he was grabbing a football out of the air.

"Keep 'em," said Stan. "Carry on with your plan."

Pan fumbled around with the carton and finally extracted a package. He opened it, tapped out a cigarette, stuck it in the corner of his mouth and again held out his hand.

"Match," he said. Stanley hurriedly produced a box of matches.

The Top Hand slowly lit his cigarette.

"How many axes, shovels, crowbars and eight-inch spikes you got in this-here storage bin?" inquired the slow-speaking Top Hand.

"I got a wagonload of that stuff maybe," said Stanley. "But who's gonna use 'em?"

"O.K.," purred Pan. "That's all I want to know. That end's took care of."

He explained his plan.

"Now, Stan, you've got your corrals and your stampede ground near finished—and your chutes are ready for action. You're gonna announce the Stanley Dowling Anahim Lake Stampede and Frontier Celebration for say June twentieth. You'll get the word out to the boys in the Chilcotin—and plaster a lot of signs and notices all over the place—with big money for saddle bronc riding, bareback, calf roping, wild horse racing and all. You can pay off out of proceeds from money spent in your store during the stampede.

"We got just thirty days to put this deal across, Stanley."

"What about the fifty miles of road to Blackwater?" asked Stanley. "The stampede for June twentieth is a top idea for me, but what about that road?"

Now the Top Hand snorted long and loud. The noise sounded like a stud horse getting branded with a red-hot iron. He leaned forward in his chair, dropped inches of cigarette ash on Stanley's vest, and barked at him.

"Everything's easy, boy. There's just nothin' to it at all. Nothin' to anythin'. All any outfit needs is one man with brains."

Stanley wasn't impressed. He brushed the ashes off his vest.

"Shoot," he said. "Let's hear what you've got."

"O.K.," said Pan. "Stampede date settled for June twentieth. I'm gonna build five miles of road from the Home Ranch towards Anahim. It will head towards the east gap in the Itcha prairie. That first five miles is easy slashing. It will take us less than a week to swamp out. That's the come-along, Stanley. That's the bait. Ridin' up the road for a mile or two anyone can see that the road has been cut out all the way to Stanley Dowling's.

"I'm gonna paint up posters and tack them in Kluskus, Ulgatcho, and the woods along the trails."

Now the Top Hand looked shrewdly at Dowling.

"Here's the meat of the deal," he said. He dropped his cigarette butt on the floor and tramped it out with his heel.

"I've already been around and told all the Indians that the Frontier Company was puttin' a road through to Anahim and that the road would be ready for travel in time for everyone to drive out in their wagon to that there celebration.

"The celebration is set for June twentieth. My posters is gonna say—Anahim Lake Stampede—June tenth. The Indians will all start in plenty of time. I'll furnish free a couple of axes, shovels, and eight-inch spikes for corduroy over the worst of the muskegs to each and every wagon of Indians that starts up the road. Tom Baptiste and Jimmy John, who's comin' up from the Batnuni, is gonna be in on the deal. They'll lead the procession with their wagons and families.

"Those boys will act dumb because they're gettin' paid double foreman's wages for the whole trip. They'll act real surprised when they come to the end of the five-mile road with forty-five miles still to go. But they're gonna tell everybody that it will be easy to make the rest of the road through to Anahim Lake if all the men get to work with axes and shovels.

"When the big crew of Indians gets as far as the top of the Itcha Mountains, where Tom Baptiste and I have already picked the route and blazed it out, the Indians is gonna start hollerin' that they've missed the stampede, it's too late to go on. Then Jimmy, who knows the date is the twentieth instead of the tenth, will ride out on a scouting trip. He returns in a mighty short time sayin' that everyone has heard that Kluskus and Ulgatcho is on its way, and done postponed the stampede to the twentieth.

"Then those Indians is gonna have to work like hell to build the rest of the road through to get their wagons over it in time for the twentieth. I figure that all it will cost you is the axes, shovels, spikes for punchen, crowbars, a few chains for pulling out rocks—and it will cost me about sixty dollars for double wages to Jimmy John and Tom Baptiste.

"The Blackwater boys will travel over the new road in time to get to the stampede when we're sure the Indians have slashed her through. We'll all be at the stampede, and everybody's happy 'cause they've all worked like hell to get there."

Stanley let out a war whoop.

"Hurray, there's nothin' to it, boy. You got her. The financing is as good as done. Let's start lining up the axes and the hardware so you can pack it all back on that string of empty pack horses."

And so the fabulous, fifty-five-mile Blackwater to Anahim, scenic, wagon highway through the high Itcha Mountains was shoved through by the combined sweat of the Ulgatchos and the Kluskus Indians. Their tents sprang up over a wide sweep of land on the headwaters of the Dean River, 225 miles from the nearest town, on Stanley Dowling's Stampede Grounds—and as Pan had predicted, everybody, even the tired and hand-blistered Indians were happy.

The Kluskus Chief's son, Peter Morris, summed the whole deal up: "That Pan Phillip, you gotta watch 'em all the time. He smart man, that Pan."

It was now possible for Pan to drive his freight wagon from Home Ranch to Dowling's, throw on a load, and land back at Home Ranch in four to five days.

It was about the time of the road-building deal that Pan and I received instructions from the company officials in New York to cut our food bills down to essentials. The approved list contained the following items: dried beans, dried prunes, rice, macaroni, flour, sugar, Roger's Golden Cooking Syrup for ranch hands, stock salt for cattle, and Bell's Medical Wonder for medicine for horses, cattle, and men.

Back here in the B.C. jungles at that time, the average workman's wage was thirty dollars a month. A man slugged away for an average of thirteen hours a day, seven days a week. Pan and I, as managers and acting foremen of our respective units, were only receiving fifty dollars a month.

I compared these conditions with outside labor and found that the average wage in other industries was around ninety dollars a month and this was for only eight hours a day, for less than six days a week. The conclusion I drew was that the hard-working cowhands were underpaid, but the price received then for beef could not support a higher wage rate. Therefore, I figured that we should supply the most wholesome food that could be freighted into the company holdings without breaking us.

When I received the grub-list instructions from the East, I immediately hooked up a team, drove from Batnuni to Vanderhoof, and out of sheer orneriness, loaded the wagon box down with everything from strawberry jam and sides of bacon to canned fruit and vegetables and tomato ketchup.

I had never gone all out this way on ranch food before, but the thought of the grub-filled Easterners suggesting that we cut down our food to fifteen cents per man per day filled me with a blind rage. I was so mad I entered the Vanderhoof liquor store, run by a former cowhand and rancher, George Steele, and bought two cases of beer for the Batnuni boys and myself, and charged them to the company.

The gang that worked for me never did know why I suddenly started turning out the fanciest food in the country, and strangely enough, outside of the usual, and I must say justifiable, kicks about spending too much money otherwise that reached me from the East, nothing was ever said about my mounting food bills.

When Pan received the same food instructions as I, he sent off the following note:

Frontier Cattle Company,
1 Wall Street, New York.

Received your letter and inform you that I will
follow out your instructions—rice, beans, macaroni, stock salt, Bell's Medicine. Come out and pay
us a visit.

Regards,
Pan

A visit Pan and I made with Paul Krestenuk at his
Nazko trading post changed the whole economy of the
Home Ranch. The price of beef was so low, wages so
inadequate, and Pan's grub so terrible, he ended up with
an almost all-Indian crew.

When Pan scrutinized the bill Paul handed him for
some food and hardware we had just purchased, Pan
studied the invoice for a moment, his face slowly turning
from red to purple, and then he exploded.

"What the hell!" he screamed in a high-pitched
voice. "I'm not gonna pay this. You're robbin' me, Paul.
You claimed you were givin' me a bargain on all this
junk you can't sell to anybody else. Bargain! What ya
mean by bargain? You've charged me full retail price for
the whole works."

"One percent," said Paul in a firm yet conciliatory
voice. "All I ever make on any sale or transaction is one
percent profit."

Pan calmed down for a moment.

"One percent profit? You'd go flat broke if you only
made one percent."

"I never have," answered Krestenuk. "I've made a good living by not overcharging my customers. One percent profit on every item isn't much but it's enough to keep the wolf from the door."

"You're crazy," snapped Pan. "What do you mean by one percent? You're making more than that on my wagonload and on every damn thing you sell in this store."

"No, I'm not," said Paul. "It's just simple arithmetic. I buy something for one dollar—I sell it for two dollar. One percent profit. That's all."

The Top Hand was so impressed with Paul Krestenuk's simple arithmetic that he went Paul one better. He opened his famous Itcha Mountain Emporium, where the Top Hand boosted the net profits from one to two and finally to three percent profit—and at times I suspect more.

CHAPTER XVI

Moccasin Telegraph

It is hard for anyone who has not actually lived in a far-flung frontier, hundreds of miles from cars, trains, telephones and radios, to visualize the many difficulties, the terrific problems of survival, the helpless and often frustrating feeling of aloneness that confronts a pioneer on so many critical occasions—when the saddle horse is the one and only means of communication.

Every once in a while the strange phenomenon of the moccasin telegraph goes into action. There are many cases that we all know of, up here, when this weird, unexplainable system hits home faster than any relay of horses could run.

No one actually understands this moccasin telegraph system. I know two men in the country who must be gifted in the art of telepathy, another form of moccasin telegraph.

One is old Missue, who could remain in contact with his closest and most receptive relatives and friends on their distant journeys. Old Missue was able to warn his two eldest daughters and some of his old-time Kluskus friends of an approaching storm, or of a game warden, or a strange group of white men who were crossing the country. Jimmy John told me that Missue was often able to transplant himself into the person he concentrated on. Missue could see the country around his friend and feel his thoughts as he rode along the windfall trails or squatted by a lonely campfire.

Missue pounded into me the danger of riding in the dark —of getting my eyes put out by snags or sharp sticks as I rode through the jungles. Jimmy told me Missue had a premonition of a snag putting out one of my eyes, and, later, when I was riding through a bunch of windfall, of a sharp stick puncturing my other eye.

Strangely enough a short time before I first met Missue he had a similar premonition—later verified—of a young Kluskus Indian who had moved to Anahim Lake country getting an eye put out during a wild-horse chase. Missue sent an Indian scout on a fast horse to Anahim to warn the man of the impending danger with instructions to ride cautiously through the bush during the coming months. When he received Missue's message the Indian just grinned.

He was a good wild-horse man and he and Ole Nucloe and several others were just getting lined up for a horse round-up in the Sugar Loaf Mountains. It wasn't the last horse run this Indian made, but he landed in Alexis Creek with one eye missing almost before Missue's messenger had started on his return trip to the Fawnies.

Just a few years ago, Jimmy John informed me that the same Indian had disregarded Missue's next warning to stop running slicks (unbranded wild horses) and save his remaining eye. I have forgotten the Indian's name, but the Anahim Lake boys know the incident

well. The Indian lost his other eye and is now totally
blind.

Having a lot of confidence in Missue, and a strong
belief in nature and its uncanny forewarnings—I took
the old man's cautionings very seriously. I have almost
lost an eye on several bush-riding deals, and I don't doubt
I would have if I hadn't been instructed to watch myself
carefully. I still stay keenly alert for high snags and sharp
limbs on any fast windfall-cracking runs.

The other man who also had an uncanny mental connection with me was Harold Baxter's father, Rae Baxter,
who owned a country home and hobby farm at the end
of the Vanderhoof road on our trail to Batnuni.

Rae is a powerfully built, middle-sized extrovert
with a barrel chest, a happy smile and a versatile mind.
An Englishman by birth, Rae played an important role
with Superintendent Ferg Parks' jungle-breaking, Public
Works road-building program here in central and northern B.C.

For years it was a long-anticipated treat for me to end
up at the Baxter farm with my tired string of pack horses,
on my freight trips from Batnuni Ranch to Vanderhoof.
It took me from three to four days to reach the Baxters'.
On the evening of the day that I pulled out of Batnuni,
Rae would casually inform his family that Rich had left
Batnuni that morning and was camping at Marvin Lake
that evening. "Rich is fishing right now from a raft in the
lake, but he's having no luck."

Everybody would give Rae the horse laugh at first.
But three days later, as I swung into their yard, a special
roast beef or duck or chicken dinner would be simmering

on the stove, and a dance planned in my honor for the evening.

Over a period of six years, Rae Baxter didn't miss once. Certainly this could not have been guesswork on his part. I usually made silent bets with myself as I swung off at the junction towards Vanderhoof whether Rae was going to miss out this time—and then again, as I rode along the trail, I wondered what Rae was doing at the time, how the horse-feed situation was at the Baxters', and so on. Quite possibly Rae picked these thoughts up. I don't know.

One of the strangest moccasin-telegraph cases I know of happened late in the winter of the Frozen Drive, when Pan suddenly decided to pay me a visit at Batnuni.

It is quite possible that Missue had something to do with catching that spontaneous action out of the pale blue horizon and mentally relaying the situation to Peter Morris. Anyway the thought waves must have got through, but the results didn't work out to plan; and I ended up in a trap that was intended for Pan.

The day before Pan left the Home Ranch I was at the Pan Meadow helping Simrose erect a chute and squeeze for the purpose of dehorning some yearlings the coming spring.

The contracting crew of Peter Morris, the Alec boys, their wives, and the Zalowie Georges had been slashing out a short-cut, saddle-horse trail through the burn and windfall country from the Blackwater crossing to the Batnuni, reducing the distance from fifty-five or sixty miles to a mere twenty-eight or thirty. So when I saddled up Montana and started out, I took no grub or blankets with me as I naturally expected to glide over the newly cut trail into Batnuni that same day.

The new trail branched due north from the Paul Krestenuk trail at the Blackwater crossing. I was surprised to see a large sign at the crossing, made from a coal-oil box, with Zalowie George's writing across it, tacked conspicuously to a big tree where no one could miss it. A hot iron of some kind had been used to burn on the letters. It read:

SHORT-CUT—BATNUNI 15 MILES

I couldn't understand this understatement in mileage as we had calculated that this short-cut trail would be around twenty-eight miles. I was equally surprised to see how wide and inviting the first mile of the road was.

The boys have done a super job, I thought. They've gone all out.

I had drifted across some twenty miles of the twenty-eight towards Batnuni when the road came to a sudden end in a great pile of impassable windfall. Standing trees, upturned roots, down timber, dead and alive, were piled ten feet high. This impassable barrier stretched away into dim distance.

It was about three o'clock in the afternoon. At the exact end of the trail the Indian road crew was sitting comfortably around a small campfire. For a moment I was speechless with combined anger, disappointment and surprise.

Then I barked at the crew, "What have you guys been doing? You were going to be through to Batnuni a long time ago."

The boys all jumped up—an incredulous look on their faces.

Peter sputtered. "Some kind mistake must be."

"Mistake, hell," I cracked out, as I crawled down off Montana. "You boys have been sleeping all the time instead of working. Now I got no blankets, no grub, no tobacco, it's getting dark, and I'm sure not turning around and heading back for Pan Meadow at this point."

I had shot my bolt. I tied Montana to a windfall and squatted by the fire. The young Indian women looked at me in a puzzled manner. Mrs. Zalowie George, who was one of my most respected and admired friends, tried in her best English to set me straight about the whole thing. The rest of the gang looked about as beaten as I was.

"Rich," she said, "Zalowie he tell you what's a matter. I don't talk white man much good."

Zalowie said, "Pan Phillip he leave Home Ranch. He ride for Batnuni. When he sees sign on road he's happy because he think he's gonna make fast trip—fifteen mile instead of sixty on the old road. He should have reached here two days ago. We didn't expect you. We thought Pan would ride up. We'd laugh, then hand him an axe. He would have to swing axe two, maybe three, days to get through to Batnuni.

"Then we'd say, haw haw, Pan Phillip. You fool us for forty-five mile—we work like hell for Anahim Lake Stampede. Now we think about it we fool you, Pan, for ten mile to get Batnuni." Now Zalowie's face became very grave. "But some mistake he happen." Zalowie waved his arm over his head. "Rich Hobson he come instead Pan—"

Peter cut in. "Pan Phillips, he smart man. He fool us all once more. This time Rich Hobson he get trapped by Pan. Rich he gotta axe 'em out windfall for two, maybe three, day to get home and then that Pan he ride through to Batnuni when road all done."

I pulled my gloves off and spread my hands over the trail fire. I looked at Peter Morris.

"What makes you boys think Pan's on his way to Batnuni? He's not riding down here until green-grass time. He told me that a long time ago. Where'd you fellows get that crazy idea?"

"I don't know," said Peter blankly. "Somebody he tell me. I don't know who. But I know Pan he ride saddle horse for Batnuni and we stop work so we can fool 'em."

I let out a snort. "Who is this somebody? There's no fresh horse tracks broke onto this trail. Nobody's been through here to tell you. I'm not plumb blind."

Peter stuck doggedly to his point.

"Give me an axe," I told him. "We gotta get through this mess, and somebody lead Montana over to the other cayuses, wherever they are, and stake him out with them."

The boys grabbed up axes, handed me one, and we slugged into the great windfall tangle.

We saddled up two days later, and jumped and bucked our horses across the last of the windfall barrier. Finally we broke out on the long, sloping sidehills of Batnuni Valley.

Pan rode up to the bunkhouse at noon the day after we arrived. I stood on the bunkhouse porch, looking incredulously at the grinning Top Hand.

He barked in his high, nasal, Nothin'-to-it tone of voice that the new cow trail was not too bad but that the crew had neglected to finish out the last half-mile of it.

"How's your dream girl?" drawled Pan as he snapped down off his saddle horse. "Has her hair growed down to her shoulders again or is she keepin' it bobbed up nice and short with a spit curl on her foretop?"

"Did you ride a hundred and ten miles down here to Batnuni to ask me about that girl?" I snapped at Pan, as he flipped his reins around the hitching rail. I noticed Pan was smoking a tailormade and that he had somehow managed to keep a long length of ash intact. I was on my guard at once, and was just going to give his arm a shove when Rob Striegler came limping around the corner of the house.

"Howdy, Pan." Rob waved, as he approached us. "Rich and the other Indians have been expecting you, but, by God, you fooled me."

Pan held his cigarette in his left hand and took a long, careful drag on it. I watched him fascinated.

"Hi, Rob," said Pan, a wide happy grin spreading across his face. "You look stove up, boy, what's the trouble with that hind leg of yours? You've done stifled yourself."

Rob was now within hand-flicking distance of the Top Hand.

"The Frozen Drive," he said. "My groin's never been right since." Rob cordially extended his hand.

I should have yelled a warning to Rob but I stood as if hypnotized, wondering how Pan was going to accomplish the feat of ash-bombing Rob and shaking his hand at the same time. But the Top Hand must have been practising. Before he extended his right hand towards Rob he slowly moved his cigarette hand over Rob's shoulder as though he were going to clap him on the back. Then quick as a flash Pan's right hand shot out to grip Rob's and the long ash in his left hand broke loose above Rob's open collar. It was a bull's-eye. The entire ash disappeared down the cowboy's neck. But if Rob noticed the villainous act he showed no signs of it.

"Come on to the barn, we'll take care of that pin-chested hesitator you've killed off," he said, meaning Pan's horse.

Pan's face fell—his ash trick had been wasted.

On the way to the barn I asked the Top Hand why he hadn't arrived three days before when he was supposed to.

"How do you know I was gonna be here three days ago? Nobody but me and Pinchy knew I was gonna leave Home Ranch when I did."

"Peter had a dream," I told him.

Pan and I tried to hash the whole thing out. Pan had pulled out of the Home Ranch at the exact time that Peter had warning of. But at Tom Baptiste's the Top Hand found Sedul, Tom's wife, desperately ill with some unknown ailment. She died of it, but much later. Her mother, Mrs. Long John, was down with either a para-lyzing 'flu or pneumonia, and two little Indian kids were crying their hearts out on the dirty ground between the ripped-out floor studding. The helpless women and chil-dren were without water or firewood, and their grub consisted only of frozen spuds and dried fish. Tom Baptiste had left for Nazko ten days before for a load of grub and had not returned.

The Top Hand had done what he could to keep the two sick Indian women and children alive until Tom arrived, three days later, with a supply of food, including canned milk for the tiny, half-starved children.

How did Peter Morris, the son of the chief of the Kluskus Indians, receive the moccasin-telegraph message of Pan's departure for the Batnuni? And what had guided Pan in his decision to visit me at Batnuni, and, in so doing, save the lives of Tom Baptiste's family?

CHAPTER XVII

~❦~

The Rough Edges

NATURE'S SAVAGE BATTLE to hold back her northern cattle frontier from the man in the saddle took a terrible toll of life and limb through those formative years. The people who pitted their strength and ingenuity against this wilderness deserve a place high on the list of our country's great pioneering adventurers.

Most of the boys who went through the ordeal of the Frozen Drive still carry scars from that bitter experience. Albin Simrose's jaw has given him trouble to this day. A third operation on it was performed just a few weeks ago by Dr. Ed McDonnell of Vanderhoof. Rob Striegler's groin still bothers him, but he refuses the exploratory surgery on it which a number of doctors have suggested.

Recently Pan Phillips told me that Benny Stobie is now completely blind, but because of his keen memory and his knowledge of cows and the Chilcotin country where he was raised, Benny is able to occupy a small, log-cabin office on a big ranch in the Chilcotin, and, as foreman, tell the cowboys where to go, what to do, and how to do it. Despite his handicap Benny is considered one of the best foremen in that part of the country.

Harold Baxter's hands gave him trouble for a long time, but this did not prevent him from building up a successful logging and sawmill operation which he now owns and runs with his father, Rae Baxter.

George Aitkens must have used up his reserve energy on that tough trip. 'Flu, then pneumonia, then a general crackup hit him, but he has fought his way back to health again and now owns his own ranch on the Mud River.

Art Lavington got hung up in his saddle, dragged and crashed through the windfalls on a bad horse, broke his leg in several places and tore his hip loose, and then nearly froze to death. Despite this crippling, Art still rides and operates his Nazko Ranch.

Nazko Postmaster Fred Rudin never recuperated from his heroic trip through a blizzard and sixty and seventy below temperatures to land the mail, the last of his many harrowing ordeals on the one-hundred-thirty-mile round trip through the snow-choked mountains, toting the mail with his teams and saddle horses. Fred died a short time later.

Old Loomis, a Nazko cowman, died with his boots on when his saddle horse lost its footing on an icy sidehill and crashed down on top of him. With his hip and one leg broken, Loomis couldn't crawl back up on his horse, but he got out his pocket knife and cut the cinch loose on his saddle horse to give the animal a fighting chance to rustle out during the terrific cold snap that was gripping the country. Loomis froze to death a few yards from the scene of the accident.

One of the saddest of the north country's many tragedies was that of the Letcher Harrington family. The youngest child wandered away from the house and drowned in the creek. A short time later Mrs. Harrington died after an emergency operation in the Quesnel hospital. Letcher was brokenhearted, but he carried on with his ranch and cattle —and then, one day when he was haying, a derrick pole broke on his stacking outfit and hurtled down on top of him, smashing his skull and burying him deep in the haystack. Letcher was killed instantly.

Letcher's great Indian friend, Jimmy Lick, was on the trail when his saddle horse pulled out on him. Jimmy tried to walk the twenty miles from camp to the nearest shelter, Joe Spehar's ranch, but with the old quicksilver thermometers in the area cracking the glass at nearly seventy below zero, Jimmy didn't make it. He was found frozen to death on the trail.

Ironically enough, the house-bound Cultus Johnny caught double pneumonia on one of his few trap-line forays, and died in the Fawnies.

Andy Holte of Anahim Lake country, one of the outstanding pioneers of that country, whom I described in *Grass Beyond the Mountains,* was luckier than old Loomis when his saddle horse started to buck with him and hit a patch of glare ice. Andy was thirty-five miles from his nearest neighbor when the horse up-ended with him, breaking his leg. The snuffy bronc disappeared over a rise of ground. Andy kept coming to and then passing out with the pain. A bunch of Andy's horses, his wild band, happened to be in the vicinity when the accident occurred. Andy waved his hat at the horses who became curious and trotted up to inspect him. He was making a desperate attempt to grab one of the cayuses by the mane to pull himself up onto the animal's back when a couple of Ulgatcho Indians saw the milling horses from a distance, rode up to see what was going on, and, by doing so, saved Andy's life. It took many days for the Indians and frontiersmen to relay Andy with teams and sleighs, travois and finally a truck, across the two hundred and forty miles of snow-bound back lands to the Williams Lake hospital, and then to Kamloops where Andy was hospitalized for several months.

But gradually those long, tough years slugged themselves out, and in the spring of 'forty-one, when the sun began climbing higher and higher into a warm, blue sky that settled smilingly and tolerantly above the high, interior plains and jungles of British Columbia, frontiersmen shrugged the aches and pains out of their carcasses and faced the lush green growth of grass, the fast-fattening cattle, the pounding of slick horses hitting across the range, with their usual comments to one another.

"Hell, man, that wasn't too tough. Most of us got stove up some, but what can you expect on a cow range north of Fifty-three?"

That was a characteristic understatement indulged in after an emergency or any series of them by the cowmen of B.C.'s last frontier.

And then in 'forty-one the hot sultry sun, alternating with periods of heavy wet rains, brought to life the great green range that could have fattened many times the number of cattle that roamed the area between the Itchas and the Fawnies, the Poplar Mountains and the Batnuni, the Tatuk Mountains and the Burnt Desert.

Our summer's problem was to put up enough hay to

eliminate the long, tough drives of the winters before, and, if possible, make a hay reserve large enough to increase the herd if we had a financial and manpower break. On every meadow small four-to-six-man crews dug away at a jungle of grass—wild green grass that reached away towards distant mountains that hung like blue ghosts against a dreamy, hot horizon. Every available Indian and his family from Nazko to Ulgatcho was given a hay contract, and meadows to cut, rake and stack.

During those years following the depression and during the war, the U.S. rancher was at last making a return on capital investment in his cattle and ranch, with commercial beef off grass selling from twenty to thirty cents a pound on the hoof; but not so his opposite number in Canada. Under the terms of a beef agreement with England, Canadian beef from stocker cows to steer beef off grass was held down to three to eight cents a pound on the hoof, and this ridiculously low return was maintained despite rapidly rising costs to the rancher on almost every commodity, including farm machinery, freight, gasoline, clothes, food and labor. Consequently British Columbia ranches continued on in their perpetual state of being in the red.

Now Pan's and my main objective was to "hold what we got," as Pan succinctly expressed it. This was not easy to do, and without the profit that Pan showed from his Itcha Mountain Emporium and his shrewd cattle-buying deals, the Frontier Cattle Company would have undoubtedly gone under financially. We had to do our level best to tough it out until the down-at-the-heel, busted, gusted and disgusted Canadian ranchers' annual gift to England expired and the embargo to the States was lifted.

But now cattle were fat, horses were feeling the effect of blazing sun on bluegrass, wild Reed's and wild beans, and thin, tired, lean-waisted cowhands were being forced to drop one button at the top of their pants.

With his ears always alert for bargain stuff, Pan heard that Paul Krestenuk had some Indian steers for sale. He saddled up and rode down to Paul's store at Nazko.

"I'm giving them to you for sixty dollars a head," Paul told the Top Hand. "Those Indians couldn't pay their grub bills and they needed money to go to Quesnel. They said, 'Take that bunch of wild cattle, Paul'—five-, six-, seven-year-old steers—shorthorns, biggest cattle ever sold in this country —twenty-six head."

Pan and Paul rode for two days in the lower Bazeko River country before they got a glimpse of the small herd. The heavy-built, spooky steers threw their tails and heads in the air when the boys rode out onto the clearing where they were grazing. Like a roll of thunder the mountainous animals crashed off into the windfalls. They were as fast as the men's saddle horses.

Pan realized that these wild oxen would bring top price at the Vancouver market, probably seven cents a pound—and he guessed that seventeen hundred pounds would not be far out on their average weight. Some of the biggest animals would weigh a ton. He also knew that it would take some hard riding and some tough horses to land them in the railroad cattle-loading pens in Vanderhoof.

Back at Paul's store, Pan said, "We've got the ridin'

fools who can get them out of here." He wrote out a check for the works.

Paul told Pan that the Indian cowboys of the Nazko had made three unsuccessful attempts to drive these wild steers to the railroad. Three years before, their first attempt had come the closest to paying off, but the whistle from a freight train decided that issue. Twenty tons of pounding beef hit the back trail for their home range. This experience made these behemoths smarter and even harder to handle; consequently the other two trail drives disintegrated before the animals were one day out of Nazko. The Indians were so sure that these jungle-cracking oxen could never be landed at the loading pens that they sold them to Paul cheap. Paul made his usual one percent profit when he sold them to Pan for sixty dollars apiece.

Pan sent me a message, via pony express, that he would meet me at Batnuni with his Home Ranch beef drive sometime early in October. There we would throw both of our beef herds together and push the combined drive through to Vanderhoof on the Canadian National Railroad. A casual side note informed me that he had purchased a bunch of steers in Nazko and, as they were more than one hundred miles closer to market at Batnuni than Home Ranch, would I pick them up sometime before beef drive and ease them to Batnuni beef range where we could later throw them into the main beef drive for Vanderhoof?

I read the letter to Jimmy John. A broad, knowing grin spread across his face when I came to the part about the Nazko steers.

"I know those steers for five, maybe six years," said Jimmy. "Pan, he sure a smart man. Those steers are too spooky to get out of the bush with just saddle horses. Pan's playing a joke on us. After we get fooled once, Pan he figures we go to him and ask him to help us. Pan he gonna bawl us all out and say nobody at Batnuni with any brains. Then Pan he gonna tell us some kind of easy way to get those steers to Batnuni and then to Vanderhoof."

I cut in on Jimmy: "I'll tell you what Pan's going to say to us, 'All any outfit needs is one man with brains.' Then he's going to snort like a horse, drop cigarette ashes down somebody's neck, and sure as fate he'll stick out his chest and show us a way of getting those steers out of there."

Jimmy said, "I think about it. By throwing brains of all Batnuni cowboys together, we maybe can equal one man."

And that's just exactly what we did. Jimmy John, Rob Striegler, Albin Simrose and I decided to have a high-level conference. There were no dissenting votes as to how and when and where we could meet to discuss the vital matter of how to corral those steers. This upper-echelon meeting got quickly organized at the barn. The solution to our problem was a simple one.

We drove a bunch of gentle, home-loving Batnuni cows into the Bazeko area where the wild shorthorn steers were ranging. A number of these cows were belled. In short order the big steers became acquainted and then attached to the old cows. There was nothing to cause alarm in the way we gently and sleepily followed the cows and their new-found admirers back some sixty miles to the Batnuni beef range.

George Pennoyer, accompanied by Alan Stuyvesant,

the president of the Frontier Company, arrived a short time before the round-up and the beef drive—and I packed them into the Home Ranch. As George, Pan and I were also directors of the company, we held a meeting at the Home Ranch with our assembled quorum.

It was decided among other things that, after landing the Home Ranch steers at Batnuni, Pan, George and Stuyvesant would ride on ahead of the drive to make all arrangements necessary for additional yard pens, holding corrals, sanding of cattle cars and so forth at Vanderhoof. It was also decided by three votes to one that Rob Striegler and I were fast becoming cranky, ornery, completely out of control and were obviously getting bushed, and for this reason the two of us were picked to escort the small trainload of beef to Vancouver, where we were supposed to limber up a bit and see the bright lights.

Everything clicked that summer and fall. We had an unexpectedly good round-up, the weather stayed at its sunny best, and more than a hundred mountainous stacks of wild hay stood like giant mushrooms on the meadows that dotted the four million acres of range and bush land controlled by the Frontier Company.

CHAPTER XVIII

❈

Men in the Saddle

IT WAS WELL INTO October when we strung the beef drive out of the Batnuni towards the distant town of Vanderhoof. There were nearly three hundred head of beefy Hereford steers, a good number of belled trail cows, and the twenty-six head of Paul Krestenuk's short-horn behemoths.

A horse wrangler from Alberta led the way, driving the horse cavvy ahead of him. Behind him Jimmy John's wife, Ellie, who was as efficient and capable as any man on a drive, had taken over the job of round-up cook and drove the four kitchen horses. These cayuses packed the grub, blankets and trail equipment, and strung down the trail behind the wrangler and the loose horses.

I followed Ellie and the pack horses, pushing in front of me about twelve head of leaders, including several belled cows, decoys for the wild steers. These wild ones were held in one bunch and slithered spookily along immediately behind my saddle horse. It was my job to hold my twelve leaders in front of me—keep them traveling down this narrow jungle trail under all circumstances. Any ball-up or milling-around business up in front of the mile-and-a-half line of snuffy range beef could easily cause the critters to split into the spruce and jackpine wilderness that stretched unbrokenly for hundreds of miles around us.

Among the bunch of wild oxen were several old swing-bag cows who couldn't travel too fast, were calm and stoical under strained circumstances, and would, we hoped, steady the big shorthorns down.

Jimmy John and Rob Striegler rode abreast behind group two, suspicious, hard-eyed, alert. Jimmy and Rob had the responsibility of those salty Nazko wild ones. It would take two good men to handle them, and, in case of any ball-up of my leaders, one of them would be close by to help me straighten them out.

Harold Baxter eased his horse along behind a winding line of steers that reached more than a quarter of a mile ahead of him to Rob and Jimmy. Harold had charge of close to seventy head of middle-of-the-road, two-year-old whiteface steers who should tip the Vancouver scales at a slightly better average than eleven hundred pounds on the hoof.

A quarter of a mile or more behind Harold Baxter, Sam Goodland, who owned the only ranch between Vanderhoof and the Frontier ranches, eased another seventy or more snuffy, trail-shy, two-year-old steers ahead of him, followed by half-blind Benny Stobie and his bunch. And, finally, bringing up the drag was Albin Simrose who kept an eagle eye out for any steers that might have eluded the men up in front. Contrary to popular belief, the drag rider on a large, bush-trail drive has a great responsibility, and only a top cowhand is picked for this job.

The complete drive, counting the horse cavvy, strung
out for nearly two miles. We strung the long, snaky, trail
herd past two holding grounds this first day of the drive
to take some of the snoose out of the ringy Batnuni
bunch. Panhandle's Home Ranch cattle, having already
been on the road for over two weeks, were trail hardened
and level headed.

At Marvin Lake, eighteen miles from Batnuni Ranch,
we had little trouble night herding the bunch, but we
didn't let up on our vigilance. At any moment the big
steers might spook at anything from an empty cigarette
package on the trail to a chipmunk or a grizzly bear and
crash off into the muskegs and jungles along the back trail.

The following night we camped at Sam Goodland's
layout and he threw his small bunch of two- and three-
year-old steers into the company trail herd, and at daylight we strung north towards Bear Creek camp.

With another trail day behind us, and the big steers
moving along towards market, I began to feel the first
flush of success. I knew that if we landed this particular
drive, including the twenty-six head of wild cattle in
the Vanderhoof stockyards, and safely into the cattle
cars, it would give me as trail boss, something that I
could brag about in the back-country cow camps for
years to come. Every cowhand on the drive felt that his
cattle-handling ability was at stake. We all realized that
one mistake, one wrong move, a moment's slacking off
could result in the loss of the wild shorthorn steers and
many of our whitefaces.

A top cowman, whether he's cutting out cattle, riding
range, roping, or moving a trail herd, is the one who can

anticipate any move that range cattle will make a frac-
tion of a second or more before the animal makes it—
then beat the critter to the punch. It takes years of
"knowing cows" before a cowhand can reach this pinnacle of achievement.

With the Vanderhoof stockyards only seven days away,
I had a strong feeling that if any outfit could rimrock this
drive into the cattle cars, this was the one that would do
it. For after all, this was the Frozen Drive crew. No salt-
ier, jungle-driving bunch of cowboys could be gathered
together in one single knotheaded group than this outfit
who had put over the impossible on that trail ordeal in
the winter of 1939.

When we tallied up the herd on the Bear Creek
Meadow we were one head shy. Sam Goodland quickly
made a count of his bunch and found a two-year-old
steer was missing.

"I've got to ship that bush cracker," said Sam, in his
quiet drawl. "I need the money."

Sam changed horses and trotted off into the windfalls
and jackpines on the back trail. It was eight miles back to
the Goodland Ranch. There Sam caught the steer. He
broke the animal to lead and dragged him into camp
sometime after midnight. Rob and I were squatting by
the campfire, drinking coffee after our ten-to-midnight
night-herding shift, when we heard a stick snap out in
the brush, a man's low voice, and then horse and rider
and steer loomed up out of the darkness.

After Sam had unsaddled and turned his horse loose
with the cavvy who were quietly grinding grass blades a
short distance from camp, he joined us at the campfire.

Rob looked up at the tired cowboy as he reached for the coffee pot.

"You ought to go along with Rich and me and the cattle to Vancouver, Sam. It would take all the aches out of your bones and free your carcass of all that rust. You look like you could do with some oiling up."

Sam grinned, shook his head.

"I'd sure like to. Haven't been to any city since I was a kid thirty years ago. Don't know what I'd do if I did get to a city or even a town."

"I'll tell you what we're gonna do when we crash into those bright lights," said Rob. "We're gonna buy a city suit, get all dressed up, and get away from the smell of horse for a change. We're going to take out some smart-dressed ladies. We're going to the best places. Rich and me is going to act like first-class gentlemen—no hollering, bragging, fighting or breaking up the furniture. No, sir, we're going high society."

"The first thing we're going to do when we hit Vanderhoof," I said to Sam, "is to leave the outfit on the holding grounds. Pan is supposed to take over night herding, cutting out and loading. Rob and I are heading straight for Bob Reid's hotel and take a good hot bath. I haven't rolled around in a bathtub for two years."

"Watch yourself," warned Sam. "I've read more people get killed or crippled up in bathtubs than riding broncs."

"I like horses," said Rob, "and I wouldn't leave 'em for long, but right now I'm gettin' clear of 'em. Going to get the stink of horse and cow out of my nose. I want to get cleaned up good, crawl into a city man's suit, put on a clean shirt and necktie, and see the other side of life."

"And when we get to Vancouver we're going to the best hotel in town," I pointed out to Sam. "Maybe the Vancouver, maybe the Georgia. We're going to get room service, and get the hotel valet to turn on the bath water and lay out our clothes before we dress for dinner. We'll have ice cubes and Bacardi rum and fresh limes and grenadine brought up to the room. I've got a blind date with a blond girl down there."

Rob grinned. The flickering light of the campfire made his gaunt face look drawn and bony, his eyes, deep set in their sockets, gave his face a strange, elfin look.

Rob said, "We'll have two of the snappiest glamour gals in Vancouver up in our suite for drinks before dinner—and they sure won't know that we're any connection with horses or cows."

"That's one thing you boys ought to be proud of," said Sam. "Bein' lined up with horses and cows. You don't want to ever feel ashamed of that."

I said, "Well, now, I'll tell you fellows something— there's nothing like planning the whole campaign ahead of time. Just like working out a round-up or cattle drive. Then there'll be no slip-ups down there in civilization."

Sam straightened up alongside the fire, took a long gulp from his coffee mug, set it down on a flat log and threw a dead willow root into the dying embers.

Sparks shot up into the night. A horse snorted some place out among the cavvy. There was a short pound of hoofs. Again silence settled down around us.

A cow's bawl drifted on the faint night breeze, and far in the distance, as if to remind us that we were not yet in the plush world of bright lights, cocktails and glamour girls, far to the east of us, the long wail of a wolf rose high on the night wind, then died.

None of us said anything. And then—a slow, almost inaudible wolf chorus hung for an instant, and then fell off into the wilderness void.

"We'll take in a good show," I said.

"And all the hot night spots," added Striegler.

"It don't take much to spook steers," said Sam calmly. "I can hear 'em stirrin' around, maybe gettin' up to their feet."

"Those damn shorthorns are worse than whitefaces if they're turned loose a few years," grumbled Rob, getting up. "They've heard wolf packs howling since they first hit the ground—they've been crowded by 'em. Wolves have dogged their trails as long as they can remember—but now—what happens? They get all excited up over a pack of black ones miles away—a good excuse they figure to stampede for the home range."

Sam had finished his coffee.

"We'll beat 'em to the punch," said Rob.

Then horses' hoofs pounded the ground. Baxter rode into the firelight.

"They're all on their feet," he barked. "Don't know what's the matter with them."

"O.K., Harold," I said. "Get around those loose horses quick. Shove them in the rope corral."

Jimmy John rose up out of his blankets.

"What the hell," he said. "Must be we slept all day again."

"Boots and saddles!" I yelled at the camp. "The herd's broke over the big mountain!"

Sam and Rob had melted into the darkness in the direction of the horse bells.

Now every man in the outfit was up and out of his spruce-tree shelter. Staked horses were unstaked—the loose ones thundered into the big rope corral. We saddled up and walked our horses out to the dark shadows of the standing steers. Harold's night guard partner trotted up.

"They'll be moving out of here in a hurry," he snapped.

"Hold 'em all in close, boys," I said. "If they run, mill 'em around in a circle. Keep riding on the outside of them, between the critters and the bush."

We fanned out around the dark mass of cattle. A wide sweep of meadow reached around them with some acreage between the bunch and the windfall jungle. And then—as if they were a single animal—the steers broke.

In that split second that can turn a stampede I heard horses pounding in behind us. Stobie and Simrose, with the other boys lined out behind them, had charged at an angle into a group of steers who had broken clear of the main herd and were headed for the bush.

Ropes snapped into the night—steers bawled—hoofs pounded the ground—men yelled above the tumult— and then the whole bunch were milling around in a wide circle. For perhaps five minutes we rode hard around the outside of the cattle. The pace became slower—and finally came to a halt.

"Was this supposed to be a stampede?" somebody called above the loud-breathing cattle.

"No, it's just a bunch of calves stretchin' their legs," retorted Jimmy John.

The next night at Line 44 camp, some of us grabbed off a few hours' sleep.

Rob groaned in his bed, "Boy, oh, boy, it's gonna be real fun to get clear of horses, cows and dirty clothes for a change—and swap 'em for lots of pretty, sweet-smelling women."

One day's drive out of Vanderhoof, Pennoyer, Pan and Stuyvesant rode out of the bush to meet the drive.

The near-stampede at Bear Creek had taken the keen edge off the cattle. The steers, traveling now on a car road, were moving along in small, loose-jointed bunches. They still carried their beef—but their heads were down and they didn't particularly care who rode up on them.

Pennoyer smiled at me from his horse.

"I'm passing this bottle of whiskey down the line. One good drink for every cowpoke—just one slug, remember—not half a bottle. When the outfit's in the yards you fellows can drown yourselves in the rotten stuff—but not until then."

Pan snorted and, riding up to George, relieved him of the bottle.

"I'll pass it along to the boys." Pan grinned. He eased towards me.

I was concentrating so hard on the bottle in Pan's hand that I failed to notice the long-ashed cigarette butt in his other. He held the bottle and bridle reins in his left hand. I reached for and got hold of the bottle just as Pan's right hand flicked toward my silk scarf. Two inches of ash splayed across my shirt front and down my neck.

"You ash-bombing son-of-a-b," I frothed at the Top Hand.

George Pennoyer laughed.

Pan said, "Sorry, friend, my hand slipped."

I took a long gulp out of the bottle. My saddle horse bolted, then kicked at a big steer who had prodded him in the rump with a sharp horn.

Now the Top Hand reached for the bottle, took a long gulp himself, paid no further attention to me, and rode on down past the slow-moving steers towards Rob and Jimmy John who suddenly appeared over a hill some distance in back of their wild band. I saw him light another cigarette and knew that using the bottle as a cover-up Pan was going to ash-bomb the boys right on down the line.

The next day the spooky Nazko steers, the Batnuni bunch and the Home Ranch beef were thrown into and held on a big, fenced-in meadow outside the village of Vanderhoof. The total drive distance from the Home Ranch Itcha Range was nearly two hundred and fifty miles. The steers had been on the trail for twenty-four days. The longest present-day cattle drive on the North American continent had ended—and Rob and I were headed for the bright lights.

Vanderhoof was a hundred-horse town with a total of four cars, four telephones and a population, including near-by farmers and ranchers, of nearly four hundred pleasant, easygoing souls. For a village of its size, located nearly six hundred miles by road from Vancouver, its closest city, I would say Vanderhoof was about the most cosmopolitan center I have yet to encounter. Highly-traveled, broad-humored men and women from many countries had somehow all discovered Vanderhoof, a garden spot on the quiet waters of the wide Nechako River, and driven in their stakes.

On the edge of town Mr. Smedley's huge lumber barn and livery stable reached almost to the two-pen stockyards, and, on the other side, to the main street of town.

Vanderhoof was about as western-looking as a western town in the movies. The one and only village street formed an *L* with poker-faced Jim Feeney's dilapidated-looking bank on one corner, and genial pharmacist Bob Steen's drugstore and bakery on the other.

Upon their arrival in town, Pan, Pennoyer and Stuyvesant had found that not more than fifty head of cattle or two carloads could be handled in the railroad stockyards. We needed a series of corrals, chutes and holding pens to cut out and work over, unbell and sort out this, the largest trail herd ever to hit into Vanderhoof.

Before the drive reached town, Pennoyer, Pan and Stuyvesant held a meeting with Magistrate Stephen Holmes in back of his grocery store to see what could be done. In short order the whole problem was solved.

The blacksmith, Pat Patterson, old Smedley, and many of his sons, banker Feeney, policemen Parker and Fielder, and a small army of teen-age village boys had volunteered to throw in their time and talents to erect the necessary yards. When the drive arrived on the main holding grounds a mile out of town, a corral system had been erected from Smedley's barn to the railroad track and loading zone.

"What a town," commented Pan as we rode around the herd and made a final tally of the steers. "I've never seen anything like it. This Vanderhoof bunch throws right in and works like hell for a week free of charge to help us boys out. I've never heard tell of a banker, a judge, a blacksmith and the policemen of a city grabbin' up axes, shovels and hammers and doin' the honest-to-God sweatin' work on a deal with no money comin' up, and still have a hell of a good time doin' it."

"These townfolk are smarter businessmen than they know," commented George Pennoyer. "They've sold the Frontier Company on what town to do their trading with. From now on this is going to be our headquarters town."

Stuyvesant said, "I've made arrangements at the Legion Hall to throw an open-house frontier dance tonight in appreciation of all this town has done for us."

The cattle had watered and grazed for a short time and now were bedded down under a group of poplar trees. Rob and I stabled our horses at the Smedley barn and started to walk up the front street towards Bob Reid's hotel.

"We're through with them for a while," said Rob. "Clear and free of cows and cayuses."

We didn't quite clear the bank building.

Bank manager Jim Feeney tapped on his office window. His portly manner and his gold-rimmed spectacles added to his solemn businesslike demeanor as he beckoned us to come into the bank.

I always did like bankers. They have a different effect on me than they do on most people. I usually break loose and tell them all my troubles and nearly always, before I realize the full impact of my folly, I find myself saddled with loans that would shake the average European nation—and it's always the same story, I end up working for the bank instead of myself.

Banker Feeney addressed Rob and me in a cold and firm manner.

"I want you gentlemen to see our directors' rooms. They are unusual for a town of this size."

"Glad to," I said. "Thanks a lot."

The black-haired girl behind the ledger stared at her figures. A young, pink-faced clerk flapped away at a pile of dollar bills behind the teller's cage.

Rob and I followed the banker through the bank, out the back door, up a rickety stairway, onto a platform, and into a square, dark-paneled, room where the inevitable, long-geared, well-polished hardwood table surrounded by heavy-duty, well-shined armchairs spread out before us.

Jim Feeney made no attempt at showing us around the directors' quarters as promised. We followed him into a small pantry where an old-fashioned icebox dribbled into a pan.

The banker's movements were swift and to the point.

He grabbed up an ice pick, smashed it into a chunk of ice, dropped the results in three glasses, splashed in a generous portion of Scotch whisky, and with a deft movement pressed the lever of a soda water bottle and fizzed the glasses towards their rims.

The banker handed us our glasses, solemnly raised his in the air in front of him.

"To the Frontier Cattle Company," he said simply. "Down the hatch."

Up the street Judge Holmes beckoned us from the front

door of his Cash Grocery. Rob and I both waved and then crossed the street in his direction.

"Good evening, boys," said the dignified Judge. "Step in. I want you to see the improvements we're making on our back-room warehouse."

"We're kind of dirty to walk on your clean floors," I said.

"We're going up to Reid's hotel and take a bath," added Rob.

"And get the smell of horse out of our hides," I said.

"Pooh, pooh, pooh," said Judge Holmes. "Horses, cows, grain and grass, backbone of this country. A young man should be proud to smell like a horse."

"Thank you, sir," I said.

"Yeah," said Rob. "Thanks a lot."

We followed Steve Holmes at a fast, flatfooted walk through his main store and into a high-roofed warehouse.

Looking back over my shoulder, I caught a glimpse of the window display. There must have been more than a ton of flour stacked in the window. The hundred-pound sacks reached almost to the ceiling. A few cans of beans, a bowl of yellow cheese, a gross of candles, two small, potted spruce trees and a neat row of toilet tissue rounded out the display.

Before we had a chance to refuse, our elbows were quickly bent and we held a glass of Scotch and soda in our hands, and the well-built judge, his glass raised in our direction said, "To the continued success of the Frontier Cattle Company—and to the boys who work for it."

"Thanks, Steve," said Rob.

"Thanks a lot, sir," I said.

We drank heartily. The Scotch and soda started to taste good.

"And here's to the health and good luck of Judge and Mrs. Holmes," said Rob.

We waved at the genial world-traveled magistrate and carried on towards Reid's.

"We better get up to the room in a hurry," I said. "It'll be dark pretty quick."

"We ought to go across the street and buy a suit and some clean underwear and socks," said Rob.

George Hawker's butcher shop was fading behind us. A strong arm came up from nowhere and grabbed Rob by the shoulder.

Rob turned.

"Hello, boys. Come on in for a minute."

"Howdy, George." Rob grinned. "We're going up to Bob Reid's and clean up."

"We've got to buy some clothes before the stores close," I said.

"Come on in, you fellows," George said. "It'll only take a minute. I've got a dandy cold storage room. Can hang up six beef at one time."

We were through the door and into the butcher shop. George made no attempt to show us his meat-hanging room, but he threw open the lid of a locker. Before we had a chance to look inside his hand moved so swiftly that all I saw was the bottle of Scotch.

Glasses were filled.

"To the Frontier boys, long may they live!"

George's great mop of hair, his cheery red face and his

uncontrolled enthusiasm, added to the long gulp of whisky I downed, gave me a happy feeling.

"Good luck to you, George," said Rob.

"Thanks a lot," I said.

"Happy days," said George.

On the front street again, Rob said, "I feel real good, but we got to get cleaned up. There's a dance tonight—remember."

"Bath," I said. "We're going to miss that clothing store over there."

Maynard Kerr, in spotless white grocery apron and Palm Beach jacket, was smiling at us. He walked over to us from the front of his modern groceteria market.

"Hello, boys." He beamed. "I've got something I want to show you. The first refrigerated meat-display counter in the North Country."

He grinned impishly.

"We'll traipse cow manure in and stink up your place," said Rob.

"We're going to buy some clothes next door and take a bath," I explained to Maynard, who didn't pause as we went by his display counter at a running walk. We come up short in an immaculately clean, file-rimmed office.

Our hands were filled, our glasses raised.

When we finally emerged from the grocery store we found that McGeachy's trading post was closed. Darkness had descended.

"I knew it," said Rob. "All this partying and we don't get anything done. Now we got to take a bath and slide back into dirty black underwear that could walk off by itself."

Corporal Leo Fielder of Yukon and Arctic fame, resplendent in his Mounted Police uniform, and Constable Bert Parker of the Provincial Police had entered the hotel lobby just before us. Rob introduced me, then pounded a bell on the desk.

A giant of a man, as broad as he was tall, opened a door on the far side of the hotel lobby. His face was tanned, but a slight flush shone from his cheeks. With the movements of a boxer in training, the big, cheery-faced man swung towards us. I realized at once that this was the legendary pioneer, Bob Reid. He nodded to the Constable and the Corporal and Rob.

Rob said, "Hello, Bob, this is Rich. We want to take a bath and get cleaned up for the party tonight."

"I'll take you boys right up to your room." Bob Reid grinned. "The cattle company has the entire floor rented. Also there's a present from the management waiting for you on your dresser."

"We don't know how we can thank you fellows for sweating yourselves out building those corrals for us," I said, swinging my glance from one to the other of the three men.

The massive, six-foot-two, Irish-brogued Constable Parker clapped me on the shoulder.

"Me lad," he said, "we'll be inspecting those cattle brands and it's a might easier working 'em over with a fence around 'em than down there in an open-ranged river bottom."

We invited the two officers up to our room, and with Bob Reid taking the lead, mounted the stairs to Room 3. There on the dresser was the present from the management—a nice clean crock of Johnny Walker.

A continuation of party seemed absolutely essential. Glasses were filled. Toasts to everyone's health were made.

"And down the hatch," said the Corporal.

"And God bless your ornery souls," said the Constable.

"And good prices for your beef," said the Proprietor. "Here, what's the matter with me. We've been drinking to everyone's health—and now I've forgotten to fill the glasses again."

Striegler stumbled against a chair. His glass flew in the air. The Constable caught it and handed it back to him.

"We're gonna get ourselves all cleaned up," I said.

"We've got no clothes," said Rob. "Don't know what's a matter with me—nearly lost my glass."

"Hold it steady for a moment," instructed Bob Reid, "so I can fill it up again."

"I sure like this town," I told the gathering.

"It's a good little spot," said the Proprietor.

"There's no crime in Vanderhoof," said the Corporal.

"And no criminals," said the Constable.

There was a knock on the door. It opened. Judge Holmes entered the room. He carried a bottle in one hand.

"Good evening, gentlemen," the Judge bowed. He looked sternly at Rob and myself. We were both sagging in the middle. "Your bath, boys. I've come here to see that both of you carry out your brag."

Panhandle and Stuyvesant now rushed into our midst. They were breathing hard.

"We've made the contact—we've made the contact!" yelled Stuyvesant. "The passenger train is picking up the cattle cars at midnight instead of day after tomorrow."

"Rich and Rob! Grab your luggage," yelled Pan. "We've got to inspect those brands and load right now."

"Son of a Doukhobor," groaned Rob. "We haven't got any luggage."

"You've just time enough to make it!" yelled Stuyvesant.

"Boots and saddles, you mountain bums!" barked Pan.

CHAPTER XIX

———≪∞≫———

Big City Wild Horse Drive

THREE LONG, SLEEPLESS, hard-benched, car-jolting, caboose-riding days and nights later, two very tired, dirty, unshaven, busted, gusted and disgusted cowhands stumbled exhausted out of the caboose at the Vancouver stockyards.

"Wow!" said Rob. "I never was so tired in my life. No sleep on the beef drive; no sleep, no rest in Vanderhoof."

"And," I added, "no rest or sleep in this series of hard-benched caboose beatings we've taken."

"We've at least got clear of horses for a change," said Rob.

"As soon as the cattle are unloaded we'll grab a taxi and head up town for the Georgia Hotel," I said. "Buy a suit and some clean underwear and socks, and a couple of nice, clean, white shirts on the way."

Exuberant Tom Baird, manager of the Vancouver stockyards, met us with a cheery shout and a wide smile.

Tom Baird, a distinguished-looking, middle-sized gentleman who gave the impression of being a college professor and wore gold-rimmed spectacles, was over fifty years of age at the time. He had been a cattle rancher, a horse breaker and trainer, and range-management expert.

Tom could still ride herd and throw a wicked rope. He was happily mixed up in about every kind of large horse and cattle deal going on in the British Columbia province at that time—and, I should add, he still is.

Tom was bursting with enthusiasm. He beamed upon Rob and myself. I caught the feeling that he was about to give us good news of some sort, but was holding it in until the last possible moment.

"We're going to grab a taxi and head right up town to the Georgia Hotel and take a bath," said Rob.

"Boys!" Tom grinned. "You won't have time for that. Stuyvesant is on his way over here now. He drove down here in a day and a half from Vanderhoof.

"Big things are brewing. A bunch of hot-blooded mares Jack Dubois contracted to sell to the French government are here outside of Vancouver. They were trucked down from Clinton. They're over on Sea Island now—the snappiest bunch of brood mares in British Columbia."

"What's that got to do with us?" I said.

Tom Baird grinned pleasantly at me.

"Stuyvesant figures you boys have done a job up there for the Frontier. He's pleased with the look of things— and is going to see that you get further financing."

"All we want to do is to go up to the hotel," said Rob worriedly.

Now Tom let us have it. He barked: "Stuyvesant picked up the Dubois mares and is turning them over to the company with his compliments."

I looked at Rob. His jaw was sagging.

"What's that?" he said.

"You boys are going to drive those mares into the

yards and load them up for Vanderhoof. We've just got enough time to grab a bite of supper."

This was terrific news—that Stuyvesant had recognized the great potential of the cattle company and was now prepared to raise the necessary capital to back us in the further development and stocking of our ranges. Apparently Pan and I were winning our big and seemingly hopeless gamble.

I should have let out a war whoop and thrown my hat in the air, but at this point all I wanted to do was to forget the whole thing and head for the Georgia Hotel.

At this point Stuyvesant rushed into the office, where Tom Baird had herded us.

"Big news, boys," he cheered. "Real big news. We've got the top bunch of brood mares in the country."

Tom shoved us out of the door ahead of him, the tops of his overshoes flapping against his legs.

While we gulped down a fast supper at a cheap lunch counter not far from the stockyards, I learned more about the Dubois mares.

They were of Arab and standard-bred breeding, had run wild for several generations in the pothole windfall country back of the town of Clinton in the Cariboo district of B.C. They were fast-running balls of fire and tough as iron. It took a large squad of hard-riding cowboys, after many tries and failures, to trap this wild band in the big corrals. Eighteen of the most unruly mares, a few colts and a couple of yearling stallions were cut out of the bunch and trucked to Sea Island near Vancouver, and unloaded, the first of the intended shipment to France, before she reneged on the deal.

The wild band couldn't be corralled or caught again. When the farmers on Sea Island complained to Dubois, he said the horses belonged to Tom Baird.

The farmers then got after Tom to get the mares off their properties or there would be some lawsuits. The Dubois mares had broken down miles of fences and gathered up the farmers' gentle horses until there was just one big herd running wild on Sea Island, many of the farmers' Clydes now taken over by the wild band.

Immediately after supper we drove out to see the band. They were a trim-legged, fast-looking bunch of horses if I ever saw any, but they were moving so fast we only got a glimpse of their tails.

The big question was how to get that bunch off Sea Island and into the Vancouver stockyards, where we could load them for Vanderhoof. It would cost a small fortune to get the material and build corrals.

On the way back to the stockyards Tom explained to us what he had figured out.

"We'll drive them in," he said. "I've borrowed horses and rigging from Jack Diamond's outfit, over at the race track. There'll be you two, Stuyvesant and myself riding. My son-in-law, Gil Bakey, and a fellow named Jimmy will drive the car out in front of the band. We've just got time to make the run. It will be after midnight by the time we jump them. There won't be any traffic to ball things up when they hit the highway.

"We've picked a couple of race horses out for you boys—fast horses that Diamond's using at the track for training purposes. There'll have to be two fast horses under saddle. Stuyvesant is riding the stockyard pinto and I'll fork old Red.

"Everything's lined up and ready to go. We'll ride out to Sea Island. You boys don't know the route or the roads we'll be bringing them in on, so it'll be a good chance to look over the lay of the land as we amble along."

There was no doubt about it. Rob and I were trapped. Horses dogged our trail. Here we were—in the big city at last—in the one place where we could dodge the rough edges of the bush and jungle and take a hot bath, put on clean clothes, and as Rob said, see the other side of life for a change—where, maybe, I could find my blond dream girl—and what had we run up against? The only bunch of wild horses within a radius of nearly three hundred miles.

We had tackled a much tougher deal than any of us realized. We were about to drive a bunch of panicky wild horses, who could run as fast as race horses, through paved streets and across high bridges into a large city. The chance of horses and men landing safely in the stockyards was mighty slim.

Tom was optimistic about the whole thing. The way he had it pictured, the deal looked easy, but Rob and I had our doubts.

For one thing, we'd have to cut out the farmers' workhorses, on the dark island with no corrals, before we started. However, as Tom pointed out, the bunch of Dubois hotbloods would move so fast that the farmers' Clydes would automatically cut themselves out of the herd before the Dubois bunch even got warmed up.

The stockyards were all lit up when we arrived back from our little jaunt.

I took one look at the outfit Tom had lined up for me and started to yell. I pointed out that I had only ridden a pancake saddle a couple of times in my life, and then with very little success, and that this combination flat saddle and army monstrosity was even worse. Tom told me I'd get used to it.

I picked up one of the Thoroughbred's front feet.

"Look at these wore-down steel plates," I snapped at Tom. "What kind of an outfit is this race-track bunch anyway? They don't even keep their horses shod. If there's any fog or the pavement gets wet and slippery, this horse won't be able to stand up."

Tom looked thoughtfully at the steel race-horse plates worn down as thin as paper.

"Well," he said, "those Dubois mares we're running are barefooted, and if they can stay on their feet, I guess this old boy can too."

Then I examined the bit in the horse's mouth. It was a straight bar bit, and after trying to neck-rein the race horse about, I found he was not only cold-jawed, but had obviously never heard of neck-reining. I could, however, pull him around by using one bridle rein at a time, the same way you'd drive a team.

I pointed this out to Tom, but he didn't think I'd have to do much turning or cutting. As for the cold jaw, he admitted I might have a little trouble holding him in with the straight bit once he got hot.

"But then," said Tom, "you probably won't have to hold him in any—these mares we're driving are awful fast."

The only thing I didn't kick to Tom about was the horse himself. He was a long, deep-chested bay with a wide forehead and an intelligent eye. He stood at least sixteen hands, had good legs and pasterns under him, and anybody could see at a glance that he could run.

Rob Striegler had about the same kicks as I had over his outfit, while Stuyvesant, the old-time polo player and steeplechase rider, had drawn the only cutting horse and stock saddle in the bunch.

It was a ten-mile ride out to where the horses were banded up. We were about three miles out in the suburbs when Tom sprang another good one on us. He and Stuyvesant had been talking about some prize Thoroughbred gelding that Stuyvesant had decided to buy. I had paid little attention to the details. All I knew was that there was room in the freight car for another horse.

Tom spoke across his saddle horn to Stuyvesant.

"I was able to swing the deal all right. We're passing the sorrel's home right now. I made arrangements with the owner to stay on the alert and lead the horse out on the lawn between two and three this morning, as soon as he hears us coming—then open the gate and turn the sorrel out. It's an easy way to get the Thoroughbred down to the yards. Won't cost us a cent trucking charges. The sorrel joins the mares here on the highway, a short three-mile run and into the yards we go, just like that."

"You're crazy," I cracked at Tom. "When that sorrel gelding breaks into the bunch, if we ever get this far with them, those mares will split in every direction."

"I belled him," said Tom. "If he breaks away with any mares we can follow the horse bell."

"Good God," Rob groaned.

Stuyvesant and Tom grinned at each other.

At every block street lights flickered. I wondered how the bunch of wild mares would take to the street lights and lampposts, the clotheslines, and the buildings fronting the street.

"This is the craziest scheme we ever got ourselves into," I said to Rob.

Rob didn't look very happy.

"Some break in training this is," he said.

Gil Bakey and Jimmy were waiting for us at the battered wooden bridge where the paved highway ended and the farmers' sea-front pastures began. They had turned off their car lights and were listening for horse noises.

Our horses shied at the dark apparition of the car. We all dismounted and had a cup of thermos coffee. As we stood around the car in the dark I asked Tom why there wasn't a shot of rum in the coffee.

"It would kind of warm us up and put us in the right mood for the night run through the streets of Vancouver."

"And," Rob added, "make us forget that we're seeing the bright lights from the hurricane deck of a race horse."

I could see Tom grinning at us in the dark. He said we would all get a draught of rum when the horses were safely in the stockyards. He and Stuyvesant had arranged for that, but the refreshments were back there at the yards, not here on the job.

"Now," said Tom, "here are the instructions. Bakey and Jimmy move the car off there beyond the bridge. Lights off and no noise. They wait until they hear the horses coming. There should be plenty of time to start the car and cross the bridge well in front of the horses and proceed down the road toward Vancouver, keeping the car about a hundred and fifty yards in front of the herd.

"The big thing," he told Bakey and Jimmy, "is to get out in front of the running horses no matter what happens.

"Striegler and I will ride up the flats and jump the bunch out on the dunes or wherever we find them, and get them headed this way toward the bridge. Hobson and Stuyvesant, wait here by the bridge and be ready to spring into action fast. Before you see the whites of the lead mares' eyes, ride like hell for that bridge—the mares will follow. Bakey will be out in front of you with the car. Then with Rob Striegler and me coming in behind them, we'll pound them comfortably into Vancouver and the stockyards.

"It's after one," said Tom. "Every man to his place, but I'm warning everyone here, you've got to get out there and ride, and that also means you, Bakey and Jimmy, in the car."

Tom and Rob rode off into the night and the Sea Island pasture. Bakey and Jimmy started up the roadster and drove to position, turned off the lights and the motor, and Stuyvesant and I rode across the grass and the sand dunes to look things over.

It was a quiet night. A gentle fog was slowly settling down around us. Even the slightest noise seemed magnified. I heard the distant squawk of a sea gull far off in the darkened dunes. A rabbit scurried out of a low patch of brush and bounced off in the gloom. He made enough noise to startle the horses.

I pulled my horse around to miss a deep hole. He didn't turn fast enough—started to tumble into the hole, then made a hard jump across it. He made the far side of the pit all right, but my strange-shaped saddle and I landed down in the bottom of the hole.

Stuyvesant called, "Are you all right? Did you get hurt?"

"Catch that horse," I yelled. "Quick!"

Stuyvesant grabbed the animal's dragging reins and led him over to the hole. I threw the saddle up on the bank and crawled up. My back felt as if it were out of joint. Stuyvesant held my horse.

I lit a match and looked at the rigging.

"Broken cinch," I grunted. "Gotta move fast."

"I hate to say anything," said Stuyvesant calmly, "but I can hear the faint pound of horses' hoofs."

I didn't answer. I made the fastest repair job on record. There was a foot or so of extra latigo strap. I let out the strap, tied the cinch together with a square knot, threw the saddle on the horse's back.

"I hear them coming," said Stuyvesant. "They're moving fast. Got to get going."

Now I cinched up hard on the rigging, made a safety tuck with the end of the latigo strap, grabbed the reins from Stuyvesant and sprung up hard into the saddle.

Hoofs pounded. The black mass of running horses was upon us. My Thoroughbred started to run. We were in the middle of the loud-breathing, wildly running band. I caught a last glimpse of Stuyvesant and the pinto engulfed in the darker color of the whizzing, plunging bodies of the wild band.

Now we were out on the bridge. Eleven of the fastest mares had beat my Thoroughbred across it. I glanced quickly over my shoulder.

A motor sputtered. Car lights came on. The hot-bloods had shot out of the dunes so suddenly that the car was left behind.

Now we were across the bridge and onto the paved highway. The clack and rattle of hoofs on hard pavement sounded like an army of jackhammers. The hard-running mares, never before having felt or heard the noise of hoofs on pavement, panicked into the fastest run of their lives.

My Thoroughbred fought into his bit. I let him have his head. He stretched out low to the road. There was nothing in front of us now. We raced towards Vancouver.

Those leaders were the fastest bunch of horses I've ever come in contact with. For the first quarter they pulled out and away from the Thoroughbred. The other mares and the colts were at our flanks, then gradually falling back. The black, clattering mass of horses in the drag showed up against Bakey's car lights. Then gradually the works, car lights and all, vanished in the darkened distance behind us.

My saddle horse had gained but little on the flying mares.

They may slow down if I'm not pressing them, I thought.

I looked over my shoulder again. We were now completely alone on the highway.

I looked ahead. Far in front of us loomed the lights of Vancouver. The road was a high fill, topped with asphalt. A ten- to twelve-foot ditch half filled with muddy water flanked each side. It was as if we were riding along the top of a moat.

Man, oh, man, I thought. If a car comes along now from the opposite direction—we've had it.

The Thoroughbred was filling the gap between the mares and us. We were perhaps six paces behind the nearest ones.

Now we shot past a side road.

If I could just turn this bunch into a side road, I thought, then at least we wouldn't end up in the center of town.

A lead mare slipped on the pavement. She went down. We were traveling at terrific speed. Another mare fell over her. Three or four more piled up on top of them. I sawed my saddle horse down with everything I had in both arms and shoulders and in my back. We slid sideways.

I saw the pavement coming up to meet us—then a great pile of writhing, sliding, groaning bodies around us—and now somehow the bunch of us were straightened out and away again. The wind rushed by. The loud clack-clack of hoofs on hard pavement echoed out into the night.

The first of the street lamps flashed by—and then another. A side road loomed ahead. Two mares up in front crashed into each other, lost their footing—went down. For a fleeting moment it looked like the whole works were piled high in a great sliding shuffle of bodies. I was on top of them. Since there was nothing else I could do, I gave the Thoroughbred everything I had, leaned far over his neck. He sprang with a terrific force high in the air.

We hit the pavement beyond with a sickening thud— slid forward—then sideways. The jar had badly shaken the grand old Thoroughbred. I could feel him trembling under me.

I looked back. We had jumped over the rolling bodies

of eight mares. The other three were on their feet in front of us flashing on towards Vancouver.

Now I swung the Throughbred around and smashed the eight mares behind me into the side road. We raced past farmhouses, barns along a secondary road of dirt. Some place up ahead twin airplane motors spat into the dark. I remembered Tom telling me about an airport. A low modern building suddenly came up out of the dark, with a neat woven wire fence surrounding it. The layout had all the earmarks of the airport night-club. We flashed by.

Then we were on a long, narrow, obviously condemned plank bridge perhaps four and a half feet wide. Some twenty feet below us the sickly gleam of muddy water stared up—and then I saw the high gate blocking off the far end of the bridge. I couldn't turn back now. There wasn't room. The deep chasm below me looked nightmarish. The boards bent beneath the weight of the horses.

The mares were now trapped on the bridge between us and the gate. I pulled the Thoroughbred down to a standstill. The mares had reached the gate, swung around, and were moving cautiously back single file.

It was one of the strangest moves I've ever seen wild horses make, but those mares were really smart. They edged along towards us as though they were walking on eggshells. My saddle horse stood stockstill. Sweat and lather dripped. His breathing was loud and hollow.

The line of mares came abreast of us, the body of the lead mare touched my leg as she eased past us. I held my breath. There was just exactly enough room for her to squeeze by. All eight of them made it.

Now I eased my saddle horse to the gate and turned about and recrossed the planks. I overtook the wild band in front of the darkened night-club where they had stopped and were milling undecidedly about. With a fast dash I shoved them into the enclosure in front of the building. The fence was only three feet high but the mares were wheezing for air after their phenomenal seven-mile run. They stood together in a bunch facing my saddle horse who filled the opening in the fence.

The twin motors of the plane died down in the night. As the loud-breathing bunch of us gradually got our wind, the dark became tensely silent.

And then—far in the distance—I heard the rattle of hoofs on the paved highway. It was one of those strange strokes of good luck, and another evidence of the uncanny ability of wild horses to sense the whereabouts of their fellow herd members.

The clatter of hoofs on pavement suddenly stopped, and a short time later the three mares that had beat me out came loping up the dirt road towards the night-club. I got around them and pushed them behind the fence with the rest.

No doubt something had spooked the animals on the highway enough to turn them about. At the side road they had picked up the tracks and still lingering scent of their fellows and here they were all back together again.

I don't know how long I held the bunch on the night-club lawn. It seemed like hours. My arms ached from the shoulders down. My back was numb, and that damn freak of a saddle had really fixed my legs. They kept cramping up on me, my toes, calves and thighs—and I'm sure my head must have had a cramp too.

I couldn't figure out what to do next. Just hold those mares on the night-club lawn until they got their strength back and jumped the low fence and headed for points unknown?

I kept looking off towards the highway wondering what had happened to the rest of the gang. Only one car light showed up all night, and a couple of cars went by as dawn was breaking.

I talked with the mares—told them about the great bunch-grass range that was waiting, green and tender for them far up north. I asked them to give me a break and not try to jump the fence and crash away through the darkened city again. They finally quit stalking along the edge of the fence looking for a way out.

I'll never forget the first sight I got of Tom Baird. Daylight had come, and with it a fog, a thin wet one.

I was as stiff as a board—and cold. Hadn't crawled off my horse since we flashed out of Sea Island. I heard a car coming up on us out of the fog. The airplane motors started up behind me again. The mares tensed—their ears and tails stood straight in the air, they started to mill.

The bunch of us will be away again in a minute, I thought.

And then here was the car with Tom at the wheel. He was hatless and covered with mud. It ran down his face. He was soaking wet, every bit of him except his glasses.

Tom stopped the car, leaned out the window and yelled, "Hurray!"

The mares whirled around the feeble enclosure.

"You've got 'em," he said. "Good going. Knew where I'd find you I'd find those horses. All of you together."

"Good Lord, man, what happened to you?" I said. "You must have been swimming in a mudhole."

"I have." Tom grinned. "Everything went wrong. The drag mares, Stuyvesant, Rob and I all beat the car across the bridge. Then Bakey tried to get around the bunch of us with the car and the only person he cleared was Stuyvesant. My saddle horse lost his footing and slid off the road and into the ditch.

"We rolled around in the water. I had one foot hung up in the stirrup, but I didn't let go my reins. Swallowed a lot of water—then Stuyvesant rode up. He's got what it takes—that Stuyvesant fellow. He may be an Eastern dude, but he's got the guts and he didn't get rattled. He grabbed old Red by the head, and held on. I got clear and crawled back on the road. I threw my bad knee out again in the mix-up.

"Then I jumped in the car with Bakey and Jimmy and told Stuyvesant to lead my horse down the highway while we drove around the country in search of you and Rob and the wild band.

"Everything's under control now," said Tom. "Rob's out on Lulu Island. His bunch hit a side road—ran eight miles before he got them turned into a plowed field. He's holding them there now."

Tom snapped to attention.

"We've got to get moving," he said. "Everything will be a cinch from here on in."

"It sure looks like it," I said. "Just make room for me in that car. I'll turn this long-geared, cold-jawed old pile driver loose with these mares, and we'll leave the works right here at the night-club. It'll be a cinch all right. You and I will drive right smartly down to the stockyards and drink up those hot rums you've been talking about—and then I'm grabbing a taxi for the Georgia Hotel."

Tom turned the car about, then called to me:

"Inside of an hour you'll hear the horses' hoofs on the pavement. As soon as you do, get in back of those mares and pound 'em for the highway. If the timing is right, we'll all meet together there, and into the stockyards we go. So long."

Tom drove away. I cursed.

Finally I heard the crack and clatter of hoofs rising up out of the foggy distance. I let the mares out and we flew north toward the highway. The two bunches came together at the intersection. Tom Baird was up in the saddle again and I saw Rob and Stuyvesant behind the other band.

"The car's out in front this time!" called Tom. "Everything's under control."

I swung the Thoroughbred in beside Tom in the lead. Tom leaned over his saddle horn.

"That sorrel Thoroughbred Stuyvesant bought is just up ahead. We'll try and hold these mares back to a lope until he gets through the gate."

"Jehovah's Witnesses!" I snapped at Tom. "Forget that horse. Leave him where he is."

I didn't have time to say more. We were passing houses now. One sat farther back than the rest. There was a lawn in front of it. A four-strand barbed-wire fence separated it from the highway.

A sorrel apparition swished suddenly across that lawn. There was the high wild zing of wire. The sorrel was with us—one strand of the wire he had run through caught in his tail.

"Ye gods," I yelled.

The sorrel circled through the running mares behind us—the wire singing around the bunch of us.

Tom yelled at Stuyvesant and Rob, "Stay on your horses for God's sake—ride like hell."

Now we were in people's backyards. Clotheslines kept coming for our heads and necks.

People were yelling—women and children were springing for shelter—picket fences crashing. The wild clang of a horse bell mingled with the high snorts and loud breathing of horses and the pounding of hoofs.

"We're taking a back route!" yelled Tom as he ducked under another clothesline.

And then just as suddenly as the bunch of us had left the highway for the backyards of Vancouver, we were back on it again—and then we were on the high bridge above the railroad tracks—and here came the trucks, moving towards us through the fog at terrific speed. We were halfway across the bridge now. Far below us the railroad tracks looked like they were in the bottom of the Grand Canyon.

Brakes squealed—trucks slid sideways—horses slipped— skidded—piled up on top of one another. Horses squealed, whinnied, groaned. Men cursed. My Thoroughbred plunged to the guard rail to avoid a truck. We've had it—we're going over the rail, I thought. I kicked my feet clear of the stirrups and threw them high over the horse's back—

I felt the terrific impact as the horse under me crashed against the guard rail. I saw blood and hair and hide streaking along its splintering surface. We were clear again. Everything was moving so fast the mind couldn't keep up with it.

And here came the great gates of the stockyards. They were right in front of us now. Cattle buyers, a great crowd of people, a line of cars blocked the road. Beside me Tom was yelling—people were waving their arms—and now the gates swung shut behind us.

"Wowie! Wowie!" yelled Tom. "We're all here! Men and horses—here in the stockyards. What did I tell you guys? —it was just like that." Tom snapped his fingers. "And we made it just in time to load them on the fast freight for Vanderhoof!"

The wild mares, the colts and the sorrel gelding, white-lathered, skinned-up, blowing like porpoises, milled around the loading yard in a close-knit, confused little knot, coldly eyeing the large crowd of spectators who jammed the fences.

A yearling stallion staggered to the loading chute, swayed unsteadily and fell over dead.

Tom Baird barked at Rob and me: "Don't let that worry you. We're loading the rest of them now—and I'm warning you boys—don't give that wild band one minute's rest when you unload at Vanderhoof or you'll never get them through those jungles. Pound them on the tail for Batnuni!"

Rob and I crashed the wild band through to the Batnuni range all right, but I never saw my blond girl— and we never got our bath.

CHAPTER XX

❦

Enter the Ladies

THE FRONTIER CATTLE COMPANY had by now survived the worst of the war years. The price of beef was up, the United States had joined the conflict and one could begin to see ahead. But many problems and much hard work remained.

In 1942 on one of my pack-horse trips to Vanderhoof for mail and supplies I was delighted to receive a letter from my mother in New York informing me that she was entraining for the frontier town of Vanderhoof within the month. Marvin Maxwell, the Industrial and Colonization Commissioner of the Canadian National Railways, who was a friend of the family's, and had given the Frontier Company invaluable help in establishing markets, was arranging the details of Mother's five-day trip across Canada to what she considered the last outpost of the white man—Indian country, a land where moose herds and grizzly bears and fierce packs of wolves charged around on the main street of town to the beats of Indian tom-toms and the blast of cowboys' six-shooters.

I was not too surprised to learn that Mother was at last coming west to visit me. She had been threatening to do so for some time, ever since the company had a head-quarters town on a railroad, and she had heard about the interesting people who had settled in Vanderhoof.

Mother, too, had wanderlust and the craving for adventure in her blood. On both sides of her family nearly every generation had produced horsemen, adventurers and cowboys. Her grandfather Houston was a cousin of Sam Houston, who helped turn the territory of Texas into a Republic and State, and in Mother's generation, two of her brothers had reverted to type. Her younger brother, Lytle Hull, after many years in the saddle, worked his way up to the manager's job on Josh Cosden's nine-thousand-head ranch on the Texas border—and her youngest brother, Russell, could throw a wicked rope and ride with the best of them. After following the gold rush to Alaska, Russell drifted south to Texas where he replaced Lytle as foreman and cow boss at Cosden's, when Lytle became manager. Russell directed the trail herd from Texas to Florida after Cosden struck oil on his Texas empire and moved a large herd of breeding stock to a Florida swamp near Lake Okeechobee.

The Hull boys had been raised in the lap of luxury in the club atmosphere of Tuxedo Park, but with that strange faraway-lands urge gnawing at their vitals, these party-going, society-raised, private-schooled dude boys proved once again that you don't have to be raised on a ranch or on the back of a range horse to become a top hand in the cow business.

My family has always been one of great contrasts in the occupations and environments of its various members. While her brothers were sweating it out on the burning Texas ranges, Mother was being presented to society in Cannes, France, where she lived with her cousins, General Baron Athanase de Charette and his American wife, the former Antoinette Polk of Tennessee. General de Charette was the grandson of the Duc de Berry, son of King Charles X of France. He was a relative of all the Bourbons and a rabid Royalist himself.

My father had just sunk the *Merrimac* in Santiago Harbor, and as Naval Constructor, was raising and salvaging the sunken Spanish fleet which he had helped to demolish, when he met and married my mother in New York. As the wife of Rear Admiral Hobson, who later became a Congressman, it was necessary for Mother to do her share of entertaining in the capital city. After my father's death in 1937 she remained in New York among her relatives and many friends.

On her way out to Vanderhoof she went to Charleston, South Carolina, to christen the U.S. Destroyer *Hobson*. Mother said she was frightened that she might not hit the *Hobson* hard enough to dent the silver case that enclosed the champagne bottle. She held her breath and took a terrific roundhouse swing that not only showered the nearest officials but bent the silver case nearly in two. One old English naval officer gasped: "My word! She did hit it an awful whack!"

It is considered bad luck if the bottle is not broken over the ship, but in the case of the *Hobson* not even Mother's mighty whack was sufficient to protect the ship. The *Hobson* was sunk during peacetime manoeuvres with a tragic loss of life.

I realized only too well that life in a frontier town

would be a great contrast to Mother's sheltered existence in the effete parts of the world, but I also felt that she would become extremely fond of the people out here and fit into the picture with her gusto for life, her love of people, and her tremendous sense of humor.

I was right. She proved a natural for this country. She hadn't been here for a week before she rented a small house and settled down to the routine life of a small town, learning how to cook, sew and wash, in between a constant round of get-togethers with her many new friends.

That summer of 'forty-two the Canadian government had put out an urgent call for more volunteers for overseas service, but men essential to the operation of ranches and farms were frozen to their jobs. Pan and I got releases from their ranch jobs for Rob Striegler, George Pinchbeck, Baxter and others who wanted to volunteer. The men left the company employ on the explicit understanding with the recruiting officers that they would be allowed to take two months' leave to help the company put up the hay. However, those promises were not kept and the boys were whisked away, and, as usual, the company was left holding the sack, with practically no help to get up that most essential product—winter feed for the cattle.

A further setback to the company operation was the loss of Jimmy John and Albin Simrose. Jimmy John had to remain on his own layout in the Fawnies to take care of the Missue family's and his own increasing horse and cattle herds, while Simrose, who had sometime before invested and stocked a Vanderhoof farm with dual purpose shorthorns, was now forced to give this his full attention.

In the meantime I had been lone-wolfing it out at Batnuni for more than a month. Hay season was approaching. I had to get out to town and try to line up some workmen, who were, at this point in the game, almost a thing of the past. It was impossible for me to leave the outfit in the lap of the gods while I went in search of help. But one day "Father" Roland Hughes, a hunter and trapper with an iron physique, a lot of cow savvy, and a good head on his shoulders, rode into camp. Roland got his nickname of "Father" several years before when he had given the Batnuni cowboys a long sermon on the merits of avoiding wine, women and profanity. The name has followed him ever since. I left the Father in charge of operations and made a fast trip to Vanderhoof where I got in contact with Harold Dwinnell, a top cowman who had recently sold his own ranch and was drifting through the country with his wife and three young children. Harold Dwinnell accepted the Batnuni straw boss job and his wife, Lucille, agreed to cook.

Craig Forfar, a six-foot, well-matured, sixteen-year-old kid who had been packing with Skook Davidson, arrived in town to visit his mother and sister June. I knew Craig would make a good ranch hand. I hogtied him and sent him out with Dwinnell to the ranch. Once more the outfit was away. This time with a new crew.

Early the following spring, when Dwinnell and a couple of the boys and myself arrived in Vanderhoof with a string of saddle horses, and three teams and sleighs to pick up our summer's food and supplies, Mother was right on deck with a welcoming party for the cowboys. She convinced Harold Dwinnell, and myself, that she was not too old or set in her ways to come back to Batnuni with the outfit through the snow-drifted mountain passes despite the possibility of a real cold snap.

We were five days on the trail. On the mountain plateau between Vanderhoof and Batnuni the mercury dipped to thirty below. One old common-sense sleigh, a relic that we had patched up for the trip, collapsed. It was beyond repair and had to be abandoned the second day out of Vanderhoof. We distributed its load of one and a half tons on the other two sleighs.

Mother decided then and there that she would help "those poor work teams out" by removing her one hundred twenty-five pounds from sleigh to saddle horse. She had been watching the four or five loose saddle horses that I always herded along as trail companions and, of course, for use in emergencies on the trail. Stuyve had caught her eye. We could not talk her out of it. Mother had made up her mind. The last four days of the trip she and her new friend became well acquainted.

Mother, riding this formerly fast and crooked bucking horse, brought up the drag of the freight outfit, driving the loose horses ahead of her. It was something that her Eastern friends should have seen and digested—Mother riding all day and sleeping nights in a pup tent set up in the deep snow.

At the ranch she threw in to help Lucille Dwinnell with her cooking job, and the two women went riding in the afternoons or tended Lucille's newly planted and quite successful vegetable garden.

Early in the summer Mother injured the joint of her little finger washing out a pair of my Cowboy King riding pants. There was nothing that Dwinnell or I could do to straighten it out. One of Canada's great surgeons lived and practised in Quesnel, so I decided to take Mother out there to Doctor Baker.

It was an eight-day trip with a freight team to Quesnel. Old Bill Comstock, the Batnuni pioneer who still lived and trapped his line on Batnuni Lake, had been having trouble with his back and he also needed medical attention. I loaded Mother and Bill on the freight wagon, and with the Bear in the lead as usual, we struck out for Quesnel.

The first day out was a rainy one. We made the McNutt place and Mother had her first experience of sleeping in a one-room cabin with a mixed bunch of Indians and whites. Zalowie George, Felix and George Alec with their wives had also taken shelter at the cabin. We all threw our food together in a kind of communal pot. The evening was spent sitting around on the floor. We had no furniture in the McNutt cabin. The Indian women remained silent while the Indian bucks, old Bill and I reminisced about bygone days and strange adventures.

This was the beginning of Mother's friendship with the Indians. It was almost unbelievable how they understood each other. These back-country Indians do not make friends easily, particularly with white women, whom they usually consider insincere and high-toned.

In the years that followed there was nothing that the Indians wouldn't do to help Mother out. When Zalowie and Mrs. George found out that Mother would not be able to return immediately from Quesnel to Batnuni with me, they made a tentative date to pick her up in town and bring her all the way back to Batnuni in their Bennet buggy. More than a month later they carried out their generous offer.

Several days out of Batnuni I drove the team into Joe Spehar's ranch through a great haze of vicious, man-eating mosquitoes. Joe cleared out a space in one of his storage sheds for Mother to sleep. She spent quite a night with a family of pack rats as her bed partners. They ran over her bed, dragged her stockings and shoes off into their own cache, but by now Mother was used to anything. She took each new experience in her stride, and said that the tougher the trip or the more violent the experience, the more it was to her liking, as it would give her additional material to tell to her friends when she went back east.

Joe Spehar insisted on driving us in his new car the rest of the way to Quesnel. Joe's new car was a 1926 Dart. It had cost him seventeen dollars including the new parts he had bought and installed himself. We drove down in my wagon as far as Paul Krestenuk's store where Joe had parked his car.

There was no bottom in the road, and the car sank to the running boards in the center of the Nazko highway almost before it had started. Paul blew a loud blast on his bugle, whereupon the Indians turned out en masse— old women, withered old men, children, and the week-end members of the Nazko River Poker Club, a sunken and bleary-eyed group of young bucks who had been in the throes of a three-day-and-night, cut-throat poker contest in an old ramshackle cabin.

Paul attached a hundred-foot length of 1½ rope to the front axle of the car, picked up a ten-foot willow stick and hollered to the Indians and myself to take the car out of there. There must have been fifty Indians of all sizes, generations, and varying degrees of talent pulling on the rope. A large fat Indian squaw who was directly behind me kept slipping in the mud, falling against me, grabbing my shirt collar and then going into shrieks of laughter. We went down twice in a pig-pile before Paul cracked his willow switch on one side of us, and Joe Spehar roared at us from the other to stop playing and "po-o-ol."

The squaw quickly forgot the fun and we towed the car the half-mile through the reservation mud flats to high land.

How Joe ever manoeuvred the old Dart with his "set-in-his-ways" type of unfinished driving to Quesnel is beyond me and Uncle Bill Comstock.

"I'll drive thees son-of-a-beetch around those mountains easy," grunted Joe.

I sat in the front seat and, under instructions from Joe, kept an eye on the speedometer.

"Thees car is fast," Joe explained to me. "You gotta watch that clock. When I go twenty-four miles an hour tell me queeck!"

Joe's hands were glued to the wheel. When we drove around a bend in the road it was always questionable whether the car would make it as Joe didn't shift his hands. Once we reached the speed of twenty-seven miles an hour on a downhill grade.

"We're approaching thirty miles an hour," I warned Joe.

"Holt on everybody!" boomed Joe. "It's too fast—it's a runaway."

Finally Joe got the car back under control again and we bumped and clanged across the bridge and into Quesnel.

To round out her summer's activities in the Batnuni Ranch country, Mother put over her crowning achievement.

I was in the middle of round-up and couldn't accompany her over the long trail to Vanderhoof. Using Stuyve as her mount, Mother joined Louis Kohse and his wife, Mary, who trapped a distant lake called Tatuk, on their pack-horse trip to Vanderhoof to get supplies. Mother made the many-day horseback journey to town with little effort or discomfort and had many laughs along the way.

The last day of the trip, when the pack horses strung into Rae and Mrs. Baxter's farm, twelve miles out of Vanderhoof, Mother insisted she would help with the milking so that she would have an even larger story to rub in on her friends and relatives in the East. According to Rae she milked two cows and actually got milk from them, and then, despite her long, tough, saddle-horse ride across the mountains, Mother went to a homecoming celebration that night, given for her by the Holmeses, the Taylors, the Kerr families and her other Vanderhoof friends. From all I could find out Mother stayed with the party until 3:00 A.M. in the morning.

She smiles broadly when anyone suggests that she must have been saddle sore, stiff and exhausted after her quite extraordinary physical achievement.

"Why, there was just nothing to it," she says. "It was easy. I've never enjoyed anything as much in my life."

Late in October of 1943 I arrived in Vanderhoof with the beef drive to hear that Mother had bought a large, partly finished log house and the twenty or so lots that went with it.

After bedding the cattle down in the holding pasture and taking care of the horses, I left Dwinnell and the rest of the boys at Bob Reid's beer parlor and ambled across the empty lots that separated the hotel from Mother's new house and property.

Tommy Smithers, Vanderhoof's top contractor, and his crew were hard at work on the finishing job. A knotty pine interior was going into place fast. Tommy greeted me at the back door, informing me that Mother and Mrs. Bob Reid and Mrs. Holmes were out on their daily five-mile walk together.

Tommy led me over the house and showed me the plans. Counting the back and front glassed-in porches, the building was forty-eight feet long by thirty-two feet wide. We looked at the library, the big living room, the ranch-style kitchen-dining-room combination, the porches, and the bathroom that was still under construction. Upstairs, a vast side-wall network of ceiling-high cupboards surrounded three bedrooms, and a third floor consisted of further storage space.

Pat Patterson, the blacksmith, had made up an oversized drum heater which stood on iron legs in the living room.

"Your mother's not only a decorator but a real businesswoman," Tommy said as we entered the living room, where the boys were rushing about nailing up the knotty pine and the stripping. "I hear she bought this place very cheap, and she's using all these pine boards for the house's interior. It's just castaway lumber out here in Vanderhoof. Knotty pine hardly brings the price to pay for its freight east."

"You've finished the fireplace," I said, walking towards the large, obviously newly built, brick fireplace.

"Yes," said Tommy, "and your mother has used it to make the first touch of home. Pictures of your family already on the mantel."

I strolled over and, just out of habit, glanced along the shelf at the old family photographs, most of which I had seen many times before. There was a picture of my brother George in his paratrooper's uniform, his master-sergeant's stripes well displayed, a grim smile on his face. It was the last picture of him taken in Italy just before he made a lone occupational jump some place behind the German lines into the mountains of Yugoslavia. At that time we had not heard from him for the eight months since his jump. A picture of myself, taken in New York years ago, stood alongside George's. I shuddered when I looked at it. I was rigged out in a custom-built city suit, looked well-fed, extremely bored with it all, and smug. I swung my gaze disgustedly away from the horror, noted my sister Lucia in an evening dress, then my father in a naval lieutenant's uniform, taken in 1898 just after the Spanish-American war. He had moustaches and a long, black beard askew at the ends.

There was a baby picture or two, and a large photograph of Magnolia Grove, the huge, white-columned family plantation in Greensboro, Alabama—and then I was staring in wild-eyed unbelief at the dream girl!

I looked quickly away and then back.

"What the hell, it couldn't be!" I blurted, then checked myself.

Tommy Smithers looked at me and then at the row of pictures, all of them framed, except the one of the girl on the end.

The unknown, blond-haired, almond-eyed vision who had haunted my trail for five years was standing there looking out at us with a kind of wistful smile on her face and a humorous twinkle in her eyes. She was dressed in a white, open-necked shirt and jodhpurs that fitted as sleek as a mannequin's. In her right hand she held a long chain and at the end of the chain, with his smooth, well-decorated head held high in the air, glared a large Jersey bull.

I forgot myself again.

"I'll be damned," I said.

Tommy laughed.

"A lovely lass she is—but don't tell me you're looking at that bull."

"I am," I said. "A Jersey bull. A milk bull. Can you beat that one, Tommy?"

"What of it?" Tommy replied. "What's wrong with a prize Jersey bull? I'll bet it would cost a lot of cash to buy that one. But who's the beautiful young lady? Your sister?"

"I'll be damned if I know," I replied. "I've got to head now, Tommy—" I started for the door. "Which road do Mother and the ladies usually take on their walks?"

I was dazed. I couldn't believe what I had just seen. Such a coincidence couldn't be expected to stand up under the hard light of reality. But it was there on the mantel—the picture of the girl.

I raced around the corner of Reid's hotel and nearly bowled over Mother, Bob Reid, Mrs. Reid and Stuart Moore, Mrs. Reid's son by her first marriage. There were happy greetings to all, kisses to Mother, handshakes with the Reids and Stuart Moore.

"And what do you think of the house?" cried Mother, holding me by the arm.

"Just tops, Mother. It's a wonderful layout. Where did you get the picture?"

"Bob and Mrs. Reid helped me pick it out," said Mother. "I owe the idea all to them—and, of course, Tommy Smithers is doing a wonderful job."

"The picture," I said. "The girl, I mean that bull. That damned Jersey bull. Where did you get it? Who is it?"

"I've been spending ten hours a day helping the workmen out on the little details I like to do myself," Mother said happily.

"I have to shove off," said Stuart Moore. "I've got to truck a load of lumber down from the Fort. See all you people later."

Stuart stepped to his truck and started up the motor.

"Come in and have tea with us," suggested Mrs. Reid.

"No, thank you, my dear," Mother said. "I want to see how much has been done on the house."

"That picture," I said. "Who is that in the picture? Where did you get it? The girl—I mean the bull."

Bob Reid laughed.

"The best thing you can do, Rich, is to come in and have a drink and steady up your nerves. You must have had a real tough trip getting the doggies out."

"Thanks, Bob—not right now. Thanks very much though."

"We'll start walking over to the house right this minute," Mother said. "I want to try out my little teapot on the new stove."

"But that picture on the mantel, Mother—"

Mother waved to Bob and Mrs. Reid and we started walking across the lots toward the happy pounding of hammers and the clean, soothing sounds of handsaws. Mother tucked her arm in mine.

"Isn't it wonderful, my boy? Enchanting people— exhilaration in every breath of this pure, clean air. This is going to be our new home—and your brother's home. I know George is coming back and he'll be out here with us in this wonderful land."

We were walking very slowly. I was amazed at my mother's complete capitulation to my stamping grounds, a land that was part of me but was most certainly a new and strange world to her.

"That girl there on the shelf? Who is she?"

"What, dear?" said Mother. "What did you say?"

"A bull," I said. "The milk bull."

We were approaching the rear end of the house. I could see Tommy Smithers moving about the kitchen with a large Stillson wrench in his hand. He waved it at us through the window.

"Oh, of course. The picture Mrs. Hill gave me," said Mother.

George, Bill and Ken Silver, Tommy's brothers-in-law, and the backbone of his construction crew, stepped

out of the back door. The work day was over and the boys
were on their way home. At any other time this meeting
with the Silvers would have spelled a get-together to
renew old acquaintance—but not right now.

When the coast was clear I manoeuvred Mother to the
front of the mantel and pointing straight at the dream
girl's picture demanded in a loud, commanding voice:

"Who gave you the picture of that bull, Mother? Who
is she?"

For a moment this blast seemed to shake Mother up,
but she recovered sufficiently to give a blow-by-blow
description of her meeting with Colonel and Mrs. Hill.
I knew we were getting closer to my objective and gave
her my full attention.

It turned out that Colonel Cecil Hill, the
Commissioner of the Royal Canadian Mounted Police,
a giant of a man with a clipped moustache, charming,
Old World manners, a twinkle in his eyes, and a strik-
ing, military bearing, and his aide, Sergeant-Major
Young, were accompanied by Mrs. Hill on one of the
Colonel's exploratory fishing trips to unknown waters.
Vanderhoof was the center of the Colonel's operations
into northern rainbow trout areas.

Mother met the Hills at Corporal and Mrs. Leo Fielder's
house, and after the men had departed on their fishing
excursion up the one hundred-fifty-mile-long, Stuart-
Trembleur-Takla Lake chain, Mother and Mrs. Hill did a
lot of heavy tea-drinking and chatting. The two women
took a great liking to each other, and Mother was to
visit with the Hills when she went through Vancouver
the following spring. At that time they would have

returned from a trip to South America where they were planning to spend the winter. Just before she left Vanderhoof, Mrs. Hill looked at my horrible picture on Mother's dresser and handed Mother the photograph of the blond girl leading the bull, saying that when Rich came to Vancouver with his beef, she wanted him to meet the young lady in the picture who was making a big success with a herd of show cattle.

I explained to Mother that my interest in the picture in question was in the fine shape of the Jersey bull's head. That the Frontier Company needed a purebred Jersey bull to breed to a bunch of milk-giving, foster mothers that we kept to nurse our numerous orphan calves. I had the feeling that the feeble excuse to cover up my interest in the picture didn't hold much water, even with Mother who knew little about the cattle business, but she let the matter ride, and the following night I was on my back-jolting way toward Vancouver, riding in the usual old-fashioned caboose, carrying in one of my pants pockets the phone number of Mrs. Cecil Hill.

I had managed to extract the disappointing intelligence from Mother that she did not even know the name of the blond vision.

CHAPTER XXI

The Blond Girl and the Jersey Bull

TOM BAIRD WAS IN HIS USUAL exuberant state of mind when I arrived at the stockyards at ten P.M., four nights and four days after leaving Vanderhoof. It had been a long, arduous trip, what with being shunted off on side tracks to let all manner of freights slide by with war supplies, frozen fish, and God-knows-what-all between the Eastern Provinces and Prince Rupert on the coast.

I unloaded the cattle wherever I could line up feed and water for them. I walked the usual two and three miles back and forth between stations and cabooses at each of the six section points between Vanderhoof and the stockyards in Vancouver, lugging a large and what seemed to be an increasingly heavy wardrobe suitcase that contained what Mother had thoughtfully brought out from New York in the way of my suits, shoes and linen.

I had wired Tom twice enroute the approximate time of the cattle's arrival, and true to form, there, as we shunted up to the well-lighted loading yards, stood Tom, looking fit as a fiddle, full of the devil—a new pair of unbuckled, four-buckle overshoes flapping.

We unloaded the cattle, filled the water troughs and rolled down baled hay to the bunch. Tom drove me uptown. We had a midnight snack at a small restaurant and discussed present cattle prices. Of my eight cars of company beef, five consisted of well-matched two-year-olds off grass. We would get a good price for these, probably between eight and nine cents a pound. Tom told me our one carload of heifery cows, some plain heifers, a few bulls and some off-color, cull yearlings would bring us from four to six cents, but he wasn't too happy about the two carloads of fat old cows. Two and a half cents a pound would be the most we could expect to get for them. In other words it would have come pretty close to saving the company money if I had put a bullet in their heads instead of bothering to ship them.

I hung around the stockyards the following morning, and before noon had revved up enough nerve to call Mrs. Hill. I had mentally improved the story about my big interest in the Jersey bull and hoped that Mrs. Hill didn't know enough about the beef-cattle business to realize that no Hereford breeder in his right mind would contemplate having a Jersey bull within shooting distance of his range. But I had to take this chance.

I was nervous and my voice not at its best when I gave the Hill number to the operator. A short ring or two, and a woman was on the phone.

"Oh, hello," I said. "This is Rich Hobson. I've been looking at that picture of the Jersey bull and I'm very much interested in his conformation."

For a moment there was a kind of stunned silence at the other end of the line. Then the woman spoke. "I'm sorry, you must have the wrong number."

She hung up.

"Wowie," I said out loud. I had broken into a cold sweat and my face felt as if a blazing hot sun had been beating down on it. I walked to the dresser, poured myself a stiff drink, took a couple of long gulps and returned to the phone.

The same voice answered.

I mumbled something about being Rich Hobson from Vanderhoof, and that I had just arrived in Vancouver with a shipment of beef.

This time the woman's voice was not in any way hesitant. It was firm.

"I suggest you call the stockyards. This is a private residence."

I sputtered for a moment into the phone after it went dead. Something wrong some place, I thought. Been in the sticks so long I don't seem to be able to adjust to other people.

I called again. This time the woman answered the phone almost before the thing rang.

"This must be Colonel and Mrs. Hill's house," I barked. "I checked the number and I've got the phone book right in front of me. I'd like to speak to Mrs. Hill."

"Why, yes, sir, this is the Hills' residence. I'm sorry, sir, but Colonel and Mrs. Hill left a week ago for South America. They expect to be away for several months."

"Thanks," I said, and hung up.

This was a hard blow. There was now only one hope of meeting the girl, and that was through the Jersey bull.

Although I had been living, sleeping and eating with whitefaces for nearly fourteen years, and could, after a glimpse or two of even the head and shoulders of a Hereford

bull, remember his conformation enough to know him if
I ever saw him again, my acquaintance with Jerseys was
so limited that the chances of finding this particular animal were slim indeed.

Vancouver was the center of a huge dairying indus-
try. The number of Jersey herds in the milk country
between the U.S. border and Hope, B.C., was large
enough to stagger the mind. But I thought Tom Baird
might have some ideas. I grabbed a taxi and headed back
to the stockyards.

It would never do to tell Tom that I had been seeing
a blond girl in visions and dreams for five years and that
at last I had encountered her picture, leading a Jersey
bull. Tom would not only razz me to the end of my days,
but would be so skeptical of my strange delusion that he
might find it a wonderful excuse to send me all over
the map on all sorts of crazy wild-goose chases.

I finally got Tom clear of the buyers and manoeuvred
him over to an off pen.

"Tom," I said quietly, "can you give me a list of Jersey
herds in this country—prize herds—you know, blue-
ribbon outfits?"

Tom staggered back with the impact of my question.
He looked unbelievingly at me through his gold-rimmed
spectacles.

"Jerseys!" he gasped. "Milk cows! What in the name
of hell is eating you?"

I cut in quickly.

"Vanderhoof farmers, Tom—a few of the grain raisers
around there asked me to look into a prize Jersey bull for
them. They've got a bunch of scrub milk cows, and want

to go into the dairy business. You know, ship cream to Prince George or something."

Tom made a loud, bleating noise through his nose. It sounded like a sheep in distress.

"A prize Jersey bull for the north country! They'd have to keep him in bed with them so he wouldn't freeze to death in the middle of summer up there."

"But the fellows asked me to look over the Jerseys around this country and I told them I would. It's not my fault if they're crazy," I snapped.

Tom made another bleat through his high-bridged nose.

"But you, Rich. All you know is whitefaces. You couldn't pick a Jersey bull from a Guernsey."

Tom and I must have yelled back and forth at each other for another ten minutes and then we walked through the cattle pens to his office. He was shaking his head from side to side when he put in a call to Chilliwack, the model farm town of the Province, and the center of the big, milk production country, sixty-five miles northeast of Vancouver.

Tom talked with someone at Chilliwack a minute or two, and hanging up the phone, barked at me to get going in a hurry.

"You're in luck," he said. "The fall fair is still on in the town, and cattle are still being shown. If you get right up there quick you'll see the best show cattle in the country, but give me a call before you buy anything. Most of these bulls they're showing are worth a lot of money, but I don't want to see them hang one on you."

I picked up a taxi in front of the stockyards. We whizzed northwards on the craziest adventure I'd ever tackled.

I had ample time to think things over as we flew along the blacktop highway. On both sides of the road small green pastures dotted with Holsteins, Guernseys, Ayrshires and Jerseys reached away into the horizon.

I stared hard at the Jersey cows as we went by. When we were about fifteen miles from Chilliwack I began to get cold feet and almost yelled at the driver to turn back.

Just what the hell am I doing? I must be balmy in the head, I thought.

What chance was there of seeing the girl anyway? There were thousands of Jerseys and hundreds of owners scattered across this hundred-mile-wide strip of dairy country.

But one thing was certain. The dream girl did exist. The picture proved that. The strange mystery of the haunting vision had bothered me long enough. If it was at all possible I meant to clean this matter up once and for all; probably one look at her in person would take care of that, and I could go back to work again with a clear conscience and another problem and another worry off my mind.

I paid my taxi driver a staggering fare, realizing that by the time I got back to Vancouver this wild spree would have cost me a month's wages, then picked my way through a long line of cattle trucks and flashy cars to the gate of the fair grounds, where I bought an admission ticket and hurried past the big show barn to the open-air ring.

People were milling about the wide doors of the barn where I got a glimpse of Belgian horses and milk cows. The bulk of the crowd was walking in disjointed groups towards the race track. I stopped a cheery-faced farmer in bib overalls and asked him what was going on. The man pointed towards the race-track fence.

"The tail end of the show," he said. "The stock parade is coming on now over there."

"Thanks," I said, and joined the stragglers heading for the race track.

I grabbed a strategic seat down on a lower level of the bleachers, close to the track, got a program from a boy in a white uniform, and looked around me.

Country-squire types in tweeds, colorful red-jacketed Royal Canadian Mounted Police, and smartly dressed women mingled with a scattering of high-brass army officers and bib-overalled, highly scrubbed, serious farmers. I began to feel extremely self-conscious and wished that I had taken the time to change into a suit. Working around the stockyards in Vancouver I wore my everyday uniform, Cowboy King rider's pants, copper-buttoned, blue denim rider's shirt, high-heeled riding boots, stamped with the telltale stockyard manure, leather wrist cuffs to protect my wrists from hay barbs and stalks, a silk scarf to keep the hay dust and chaff from going down my neck, and my trail hat which had originally been an expensive 3X Beaver Stetson, but was now battered, old and grease stained.

As the bleachers filled up I noticed that people kept looking at me as if I were one of the circus freaks. Soon my ornery and perverse streak took over and I thought: To hell with these dudes. I glared straight ahead.

The microphone boomed out from the judge's stand. The grand parade was on.

Several well-matched Percheron teams pranced by, followed by Belgians, trotting horses, Kentuckys, a few

Arabs, and then came the cattle herds. Holsteins led the way. I listened to the announcer's descriptions of the outstanding pedigreed animals, the names of the owners and the herd names. Colonel Victor Spencer's Earlscourt Farms Hereford exhibit was the finest I had ever seen.

The announcer called: "Bert Erickson, Manager of Earlscourt Farms, is leading Royal Domino, one of America's outstanding Hereford bulls, Junior Champion, San Francisco World's Fair." Strangely enough, in a few years' time I would be buying my Rimrock Ranch herd sires from Spencer's internationally famous bull, Atok Lionheart, imported from Herefordshire, England.

An attractive girl, tall and dark-haired, was leading a blocky, low-set, whiteface heifer whose flat back was wide enough to serve a luncheon for six on. The microphone boomed that Miss Barbara Spencer was leading the Canadian Grand Champion Hereford heifer. Soon after this show, Barbara became known as the Cattle Queen of British Columbia and, indeed, of all of Canada when she took over and managed the vast Dog Creek cattle empire, and the thousands of cattle that roamed across some 375,000 acres controlled by her Chilcotin ranch.

I was becoming more and more on edge as each breed cleared the stands. Except for the universally gentle Hereford bulls, it took two men, each holding a long chain snapped to the ring in the bull's nose, to lead each farm's prize animal. Young men and women dressed in white led the heifers and cows single file behind the bull. It seemed to me that nearly every known breed of cattle had passed the grandstand before the man behind the microphone announced the entrance of the Jerseys.

I glanced quickly down at my typewritten program and saw that there were only five Jersey herds entered. A notation on the bottom of the sheet stated that these well-known herds were comprised of many of the finest show cattle in B.C. and that the animals on parade held some of the top milk production records in Canada.

From the microphone came the announcement: "A.E. Dumvill's Lindell Farms Jerseys coming up." My nerves tensed like taut telephone wires. The first Jersey bull was close now. He was fighting his head but the two lead men were handling their chains expertly. I stared hard at him. He was not the bull. His face was longer and his horns smaller. Other cows and heifers passed by.

Now Harry Reifel's Bella Vista Farm herd was lined out in front of the stands. Still not the right bull. There were no golden-haired girls. Grauer's Frasea Farms glided slowly by. My eyes lit on a slim girl, leading a calf. The announcer called out: "Miss Marjorie Antenen leading the Junior Champion Jersey bull of this show." The young lady was brown-haired with a cute, turned-up nose, but it was not the girl.

I became panicky. Only two herds left. I stood halfway up in my seat. The announcer called: "Savage Brothers. Ladner, B.C." I strained my neck in the direction of the approaching line of cattle and herdsmen. I took a deep breath as the herd bull came in close. But I grunted the air out fast. I heard the name, "Sunny Boy—Reserve Grand Champion." It was a great, proud animal of fine breeding, I saw that, but his neck seemed set into his brisket differently, and he had too much white on him.

The rest of the animals and their herdsmen moved on by. I flopped back into my seat. Perspiration broke out all over me.

"This next one is the last herd," I said to myself. "Last chance."

I reached into my pocket for my handkerchief, wiped the perspiration off my face.

The announcer blasted out: "Mrs. W.R.W. McIntosh's Moy Hall Farm Jerseys." I saw the herdsmen, the lead bull and the long line of Jerseys coming toward me. I held myself down to the bottom of my seat, reaching into my pocket for a tailormade. I started to light it but burned my finger with the match.

The men and the lead bull were almost up to me. I sucked in my breath. The bull lumbered proudly and gracefully toward the grandstand, a wild gleam in his eyes. His neck was massive and arched. I had seen him before—I could never make a mistake on this one.

The announcer called: "This bull is known as The Colonel. He is Lindell Signalman Standard—third time Grand Champion Jersey bull of this show."

I stood up in my seat. The Colonel was close to me now. I kept staring at him, half hypnotized.

Now the announcer's voice boomed across the track: "This is Moy Trixie, Grand Champion female, led by Gloria McIntosh."

I snapped my head around, staring in utter amazement at the girl leading the cow. I could have reached out and touched her wavy blond hair. It was the Dream Girl.

The bleachers were emptying and I moved with the crowd. I had to think fast. How should I approach Gloria?

To suggest that a Mrs. Cecil Hill had told my mother—
no—that angle was out. Introducing myself as the man
who's had a dream girl and it's she, would place me in the
category of a lunatic at large.

There was only one way out. I must show a great
interest in the Colonel, the Jersey bull. To prove that I
meant business I must make her an offer for him. If she
accepted, there was little doubt that I could resell him,
probably at a loss of not more than three hundred
dollars.

Now I wished Panhandle Phillips were here. He
would have an idea—he would know how to deal for
the bull. He would give me some sound advice on how
to handle this whole complex situation. But Pan was
not here.

I neared the gigantic show barn. Big crowds were
milling like undecided range cows around the front of
the building. Newspaper reporters and photographers
surrounded groups of herdsmen and show animals. Flash
bulbs were going off.

I pulled up short among a small mob of people surrounding a bull. It was the Colonel. Seven hundred
dollars was the limit. I would offer the dream girl seven
hundred dollars for him. If she accepted I could pay four
hundred dollars down and draw three hundred dollars
against company expenses. I could pay the company
back later on out of wages.

I gritted my teeth, removed my hat, took a deep
breath and stepped firmly forward to confront the dream
girl. I made a slight but what I thought was an impressive bow.

"How do you do, Miss McIntosh," I heard myself saying. "My name is Rich Hobson. My company, the Frontier Cattle Company, is interested in buying your famous bull."

I noticed how tanned the girl was, how her clear, green eyes were tinged with amber flecks, and how her short white gloves contrasted with her slim, brown arms. She extended her hand to me graciously and looked impersonally straight into my eyes.

The dream girl informed me that the Colonel was not for sale.

"But I'm willing to pay at least seven hundred dollars for him," I said.

For a moment the blond seemed stunned. She stared unbelievingly at me.

"Seven hundred dollars for Lindell Signalman Standard!" she gasped. "Not really! What did you say your name is?"

"Rich Hobson."

"Mr. Hobson, I'm sure you must be joking."

I suddenly realized that I had made a bad blunder. This is a hell of a start, I thought to myself. I'll have to straighten this out.

"I'm sorry, Miss McIntosh, if you think my offer is too low. We need a bull like that for our foster mothers up north. I'm not much of a cattle buyer when it comes to milk cow stock. Is seven hundred dollars such a terrible offer for that bull of yours?"

The blond girl stared scathingly at me.

"It's perfectly insulting to the Colonel."

She whirled on the heels of her white shoes and

walked off toward the line of trucks and cars. Several men and women drew in around her and I watched the group string out through the gate.

I stood there looking down at the turf, scratching the grass with the heel of my riding boot, feeling like an idiot.

One of the tweed-suited friends of the girl walked briskly toward me.

"Oh, just a minute there, Hobson," he said when he reached my side. "I couldn't help overhearing part of your conversation with my sister. I want to introduce myself. I'm Jack McIntosh, Gloria's brother."

The young man's face was friendly. His eyes were smiling at me through his tortoise-shell glasses. I felt instinctively at ease with him.

"Glad to meet you, Mr. McIntosh," I said. "I'm afraid I insulted your sister with a haphazard offer I made for your herd sire."

Jack's laugh was infectious. "Don't let that worry you. Baby is so proud of her home-raised bull that she's lost her sense of humor about anything connected with him. I'm much interested in the north country. I've heard about your Frontier Company ever since you Americans moved in there. There's an article in this morning's paper about your shipment of beef and the long drive you had getting them out to the railroad."

"It's a pretty long trail all right," I answered.

"The longest drive in North America, according to the paper," said Jack. He suddenly became businesslike.

"Have you got a ride back to Vancouver?" he asked. "Gloria, my brother Fraser, and a gang are all stopping off at the farm on the way back to town, but I've got to be in Vancouver at five o'clock." He glanced at his wrist watch. "I've just got an hour and thirty minutes."

"Let's get going," I cracked.

The two of us went out of the Chilliwack fair grounds at a fast running walk.

Jack led the way to a big, gray-blue Imperial Chrysler. We jumped in, slammed the doors. The motor hummed for a moment and then we burst away towards Vancouver like a rocket bomb.

Instinctively I held on to my hat. The scenery close to the road was blurred out with the speed. I pressed my feet hard to the floor and tried to be nonchalant. For a short time I wished this was Joe Spehar at the wheel of his Dart, and that I was keeping tabs on the speedometer. But Jack knew the road, there was little traffic and I soon settled back in the seat, and we started to powwow.

Jack knew more about me than I had imagined. It turned out that Mrs. Cecil Hill was his aunt. She had mentioned meeting my mother when she was on her northern trip.

Jack and I had a lot in common. He had pioneer blood in his veins, and hoped some day to establish a beef-cattle ranch somewhere in northern B.C. So far his closest acquaintance with the rugged end of pioneering had been week-ends on the McIntosh farms, and occasional boat trips up the coast to the Bella Coola country where members of his family owned salmon canneries.

Jack dropped me off at the Georgia Hotel. We agreed to meet the following day for lunch and I rode the elevator up to my floor. I slipped into a pair of shorts, went through a few light, floor exercises and a couple of rounds of shadow-boxing, then jumped under the shower.

I slipped on a white shirt that didn't look too frayed at the edges. I squeezed into a blue suit which felt a trifle small under the armpits. I hadn't worn it for eight years. I was just picking out a tie when the phone rang.

Must be Tom Baird, I thought, as I stepped quickly to the telephone table.

Jack McIntosh was on the other end of the line.

"Are you doing anything tonight?" he asked. "How about joining me for dinner at the golf club?"

"Great, Jack. Delighted. Fine."

"I'll pick you up at the hotel at seven-thirty."

I was waiting in the hotel lobby when Jack arrived. We walked out of the hotel. The doorman opened up both car doors.

I bent into the back seat and nearly fell over the dream girl. In the brief flash of light through the open car door I saw that she wore a trim little fur-edged hat, and her golden hair curled down onto the shoulders of a brown velvet suit.

CHAPTER XXII

—◆—

With the Help of the Cows

THE FOLLOWING MORNING I was at the stockyards before the gates were thrown open. Tom Baird was not far behind me. I followed him out on the ramp above the Frontier Company cattle pens.

Tom called to a couple of stockyard laborers to refill the water troughs.

"Those critters are filled up on that dry hay. Let 'em drink all they can now. You fellows ought to know better than to dry them out before the buyers arrive. Snap out of it," he yelled. "Turn on those taps."

Tom swung around. He winked slyly at me.

"Water weighs a hell of a lot," he said.

I knew by the look on his face that he had good news of some kind.

"I sold three lots of your steers to Jack Diamond's Pacific Meats for nine cents. That was the top of the market. The rest for eight cents. Not too bad, Rich. I got seven cents straight through average on your two-year-old heifers to Swift's and Burn's. That's a good price, considering you didn't dehorn them.

"Canada Packers made an offer of two cents on your two cars of cows. I turned it down. Today they'll go for two and a half cents sure. That's what you've got a commission man for—to get you the best price."

I looked down at the old company cows that filled the two pens below us. I knew everyone of them personally. We had been through good weather and bad weather, green ranges and frozen ranges together.

The cattle buyers, the packing-plant employees, the stockyard attendants who would now be in close contact with these leg-weary, trail-worn old cows, these *grandes dames* of the Itchas, the Fawnies, the Ulgatchos, would never know the story that lay behind them. They would never know that this gallant herd of old cows had trailed across the last of the free-grass cattle frontiers of our continent, more than a half-century later than their ancestors of the Chisholm Trail days. They had struggled across hundreds of miles of little-known, uncharted country; had crossed great mountain ranges and swum ice-choked rivers; they had faced wolf packs and raging Arctic blizzards. They had known the gnawing pain, the aching belly of creeping starvation on frosted bodies. These cow critters had made history.

A blanket of sadness began to settle down around me as it always did when I had to say good-bye for the last time to my trail companions. I had cursed them and praised them in turn. No cowhand ever asked for a headier, tougher, more range-adjusting bunch of old mother cows—but, now, I thought—

These soggy, loyal, old girls who have nursed so many fat happy calves are going to do one more favor for Rich Hobson.

I turned to my commission man.

"Tom," I said, "what is the very top price you can get for these cows?"

"Two and a half cents is better than market," he replied.

"Are you sure you can't squeeze three cents out of the packers?"

Tom yelled, "Three cents! Forget it."

"O.K. then," I replied. "I want four cents a pound for every head down there."

"Four cents!" Tom made a long bleat through his nose that shook the stockyard laborers out of their lethargy and startled the cows in the pens below us.

"Four cents! Man, oh, man, you'll never sell those critters until Christmas at that price."

"Fair enough, Tom," I snapped.

"Your ranch," yelled Tom. "You've got to sell those cows before you go north. You know that. You'll be stuck here in Vancouver until they're cleared."

"Four cents, Tom. Four cents on the hoof—for a great bunch of old cows."

I right-about-faced, waved an arm at the thunderstruck Tom and headed for a taxicab.

One week later Gloria stood beside Tom and me on the ramp. Tom, who had known Gloria since she was a little girl, was bawling me out over the four-cent price I had set on the old cows who stood happily chewing their cuds below us.

"And I told him it would take until Christmas to sell those pets of his at that price," he concluded.

Gloria smiled at him.

"You know, Tom," she said, "I can see Rich's point. It won't hurt those poor old cows one bit to stay here and get fed and rested up until after Rich and I are married and have to go north."

Tom's mouth flew open.

"What's that?" he said. "What did you say Gloria? You and Rich married?"

For a moment Tom shook his head from side to side, then he grabbed my hand.

"Man, oh, man!" he said. "This is one time that old cowboy saying hits the nail on the head."

I knew what he meant. I laughed.

Gloria said, "Which saying is that?"

Tom's face broke into a wide smile.

"From now on, young lady," he said, "you'll be hearing it a lot. There's nothing too good for a cowboy!"

RICHMOND P. HOBSON, JR.

GRASS
BEYOND THE
MOUNTAINS

A TRUE ACCOUNT OF DISCOVERING
THE LAST GREAT CATTLE FRONTIER

With laconic cowboy humor and the ease of a born writer, Richmond Hobson describes the life-and-death escapades, the funny and tragic incidents peopled with extraordinary frontier characters, in a true adventure that surpasses the most thrilling Wild West fiction.

In the fall of 1934, three cowhands with a dream of owning a cattle ranch made their way from peaceful Wyoming to the harsh, uncharted territory of the British Columbian interior. In conditions as challenging as any encountered by the western frontier pioneers of a hundred years earlier, the three men and their equipment-laden horses conquered the tortuous miles over narrow passes and mountain summits, hewed their first cabin from virgin timber, and attempted to carve out a space for themselves on the unforgiving landscape.

Gritty, fun, and endlessly entertaining, Hobson's story is sure to entertain country- and city-dwellers alike.

978-1-4000-2662-3

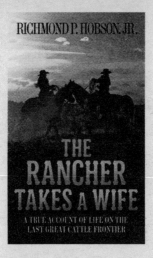

RICHMOND P. HOBSON, JR.

THE RANCHER TAKES A WIFE

A TRUE ACCOUNT OF LIFE ON THE LAST GREAT CATTLE FRONTIER

The interior of British Columbia in the early 20th century is a jungle of swamps, rivers, and grasslands. It's a vast and still barely explored wilderness, whose principal citizens are timber wolves, moose, giant grizzly bears, and the odd human being.

Into this forbidding land, Rich Hobson, Pioneer cattle rancher, brings Gloria, his city-raised bride. Her adjustment to life in the wilderness is sure to be difficult, as is her relationship with Rich and his backwoods cronies. Will Gloria find that she belongs in this strange, harsh land?

Told with wit and wisdom, Hobson recounts a wild true adventure story in the last book of his collection of survival tales. These dramatic tales are described with the humor and vivid detail that have made Hobson's books perennial favorites.

978-1-4000-2664-7